PERSPECTIVES ON ORGANIZING CRIME

Perspectives on Organizing Crime

Essays in Opposition

by

Alan A. Block

Administration of Justice,
The Pennsylvania State University

KLUWER ACADEMIC PUBLISHERS
DORDRECHT / BOSTON / LONDON

Library of Congress Cataloging-in-Publication Data

Block, Alan A.
    Organizing crime : essays in opposition / Alan A. Block.
        p.   cm.
    Includes index.
    ISBN 0-7923-1033-0 (alk. paper)
    1. Organized crime--United States--History.   I. Title.
HV6446.B57   1990
364.1'06'0973--dc20   90-19152

ISBN 0-7923-1033-0

Published by Kluwer Academic Publishers,
P.O. Box 17, 3300 AA Dordrecht, The Netherlands.

Kluwer Academic Publishers incorporates
the publishing programmes of
D. Reidel, Martinus Nijhoff, Dr W. Junk and MTP Press.

Sold and distributed in the U.S.A. and Canada
by Kluwer Academic Publishers,
101 Philip Drive, Norwell, MA 02061, U.S.A.

In all other countries, sold and distributed
by Kluwer Academic Publishers,
P.O. Box 322, 3300 AH Dordrecht, The Netherlands.

Printed on acid-free paper

Printed in The Netherlands

For Marjorie Leach and the memory of Lillyan Block.

Table of Contents

INTRODUCTION

This volume of essays is titled *Organizing Crime* because I am firmly of the opinion the gerund rather than the adjective better conveys the ceaseless activity at the core of that which is called organized crime at all other times. The subtitle, *Essays in Opposition*, was chosen for the essays share a certain quarrelsome, revisionist, similarity. One reason for this cantankerous perspective is that I have long felt that much of the social scientific literature dealing with organized crime suffers from a particular kind of ahistoricism which critically weakens arguments about the nature of organized crime.

The book's first section addresses these issues in two essays written over a decade apart. Both investigate alleged turning points in the history of organized crime used over and over again by criminologists which seem to me unsubstantiated. Thus, I argue that conclusions derived from suspect historical claims are inevitably off the mark. At the same time, I am mindful that this problem is related to the failure of historians in general to undertake serious studies of the development of criminal syndicates. Obviously, this is one reason why the criminology of organized crime is so often weak. Without the kind of research historians have brought to bear upon such phenomena as the Atlantic slave trade, the French revolution, English politics in the age of Cromwell, and so on, historically-minded criminologists have little sophisticated work to turn to. Social scientists concerned with the state and nature of organized crime typically rely on either government pronouncements of organized crime's history, which are typically self- (and state-) serving, or work from exceptionally poor, often undocumented, secondary sources.

This dearth of historical writing is one of the predominant motifs of section II, in which some aspects of the traffic in narcotics is explored. Interest in drug trafficking has exploded during the past decade, and outstanding work is now being produced dealing with issues such as illicit drug production, transportation, and distribution. Almost all this work is resolutely contemporary, however. The burden of the essays in this section is to flesh out parts of the historical background on both a micro level (in Progressive-Era New York), and on a much larger canvass exploring changes in drug production and distribution particularly in Europe from about 1925 to the mid-1930s. Both chapters address the ethnicity of traders, the structure of drug syndicates, and the impact of legislation which attempted to criminalize more and more of the world's narcotic industry prior to World War II.

The next section concentrates on organized crime's fairly recent penchant for environmental criminality. The chapters are taken from testimony before a U.S. Senate Subcommittee in 1983, and a Pennsylvania State environmental committee in February 1990. Both selections are far more personally presented *than any of the others in this collection. They represent a merging of academic interest with a commitment to environmental activism which occurred after several years of research conducted under the wing of the New York State Senate Select Committee on Crime created in the 1960s to investigate organized crime. I had the rare opportunity to work with this committee under the direction of counsel Jeremiah McKenna at the time attention was focused on organized criminals moved into the hazardous waste disposal industry. Much of my data came from committee investigations, hearings and reports. In addition, contacts outside of New York were eased by this affiliation. My long association with the committee inexorably led to an integration of research findings with an involvement in the formation of public policy.*

Part IV presents another usually hidden dimension of organized crime; its mesh with transnational political movements, intelligence services, and political murder. Two essays dealing with assassinations a decade apart form this section. One was the murder of Carlo Tresca in 1943, the other of Jesus de Galindez in 1956. The former reveals the interrelationships between Italian-American organized criminals and the Fascist party, the latter the netherworld of professional criminals, private detectives, intelligence operatives, working together in the interests of dictator Rafael Trujillo of the Dominican Republic. There is a natural affinity between American intelligence services and organized criminals which can be traced by a detailed examination of these murders. This same point is also presented in one of the chapters in Part V. In particular, it is claimed that America's various wars on drugs have been hopelessly compromised by cold war premises that transformed agencies such as the Federal Bureau of Narcotics into counterintelligence forces. Drug producers and smugglers were divided into "good" ones and "bad," depending upon their utility in the fight against communism.

One other related and disturbing ambiguity in crime control efforts is addressed in Part V. During the 1970s, the U.S. Internal Revenue Service conducted an internal war which had the consequence of making the world safer for scores of sophisticated so-called "white collar" criminals in league with organized crime and, once again, U.S. intelligence services. This essay argues that I.R.S. efforts against serious criminals (those who established important off-shore tax havens to service the wealthy whether organized criminals and drug smugglers or not), as opposed to "mom and pop" tax evaders, was crippled.

Finally, the extension of American organized crime activities abroad is considered. The illicit enterprises of a criminal cabal led by the infamous racketeer, Meyer Lansky, is tracked. Among other places, this syndicate operated in Amsterdam, The Netherlands. The question is asked, given this presence, whether the European experience with organized crime is as different from the American one as many have thought? The collection ends with a familiar refrain; without sound historical studies and a more sophisticated criminology, the essential nature of organized crime in the twentieth century (whether in the U.S., The Netherlands, Germany, or indeed anywhere else) cannot be understood.

Any scholarly work depends to a large extent on the generosity of friends. In this regard I must mention the outstanding help of my department, chaired by Daniel Maier-Katkin, which so generously supports research and writing. I especially wish to commend the labor of Senior Secretary Melody Lane who typed and shaped what follows. Two of the essays have co-authors. My wife, Marcia J. Block, co-authored Chapter Eight; professor Bruce Bullington, my friend and departmental colleague, did the same for Chapter Eleven. They both allowed me to present our work in this book. I also wish to thank the following publications for granting permission to reprint essays: Criminology; Urban Life; Journal of Social History; Contemporary Crises; Social Justice; Research in Law, Deviance, and Social Control.

PART I:

THE HISTORICAL PERSPECTIVE

Chapter 1
History and the Study of Organized Crime

Published in <u>Urban Life: A Journal of Ethnographic Research</u> (1978)

In the April 1965 volume of the <u>American Sociological Review</u>, historian Stephan Thernstrom discussed what he termed, "The Perils of Historical Naivete." Thernstrom's target was a portion of W. Lloyd Warner's work on the social structure of Newburyport, Massachusetts. Thernstrom wrote that "Warner's unwillingness to consult the historical record and his complete dependence on material susceptible to anthropological analysis-- the actions and opinions of living members of the community--served to obliterate the distinctions between the actual past and current myths of the past."(1) Thus, he concluded, social scientists have "accepted uncritically the community's legends about itself--surely the most ethnocentric of all possible views."(2) Thernstrom's comments are equally applicable to much of the social scientific work dealing with organized crime. There is a standard history of organized crime which clearly relies almost exclusively "on material susceptible to anthropological analysis--the actions and opinions of living members of the community"--most notably the alleged word of Joseph Valachi, organized crime's most famous informer.

One of the most important issues in the traditional history of organized crime concerns the origins of the so-called Cosa Nostra or as one of the latest writers, David Leon Chandler, calls it, "The Making of the Syndicate."(3) Donald Cressey wrote in "The Functions and Structure of Criminal Syndicates" that "the basic framework of the current structure of American organized crime, . . . was established as a result of a gangland war in which an alliance of Italians and Sicilians was victorious" in 1930-31.(4) The culmination of this reputed war was the execution of Mafia boss, Salvatore Maranzano, and an alleged "purge" of old-style Mafia leaders. "The successful execution of Maranzano was the signal for the planned execution of some sixty Maranzano allies, called 'Mustaches,' a reference to their traditional Sicilian ways," according to Chandler whose account is not materially different than a host of other commentators. Chandler continues:

> Details of the purge are not known, even after
> forty years. Implemented on a national scale, it
> must have required extraordinary preparation and
> communication. Each of the sixty victims must

have been kept under surveillance to establish his
daily pattern. For each of the sixty, a hit team
had to be organized and gunmen chosen who
wouldn't betray the plan. When Purge Day
arrived, the hit teams had to be delivered to their
target's area. A communication liaison must have
been worked out to relay the go-ahead message
from New York--that Maranzano had been killed-
-to each of the teams.(5)

Although Chandler cites no sources in his book, it is unequivocally clear from
the context that his source is Joseph Valachi. Chandler's story parallels the
one told by Cressey in his book, Theft of the Nation, except for the number
killed. Cressey is more conservative claiming only about forty "Italian-Sicilian
gang leaders across the country lost their lives."(6) Cressey's section on the
history of organized crime, which he properly calls a "sketch" not a history,
is acknowledged to be "based principally on the memoirs of one soldier,
Joseph Valachi."(7)

This brings us to the first major example of historical naivete which is
composed to two seemingly contradictory parts. First is the almost total
reliance on Valachi and second is the apparent disregard for Valachi's
testimony before the Senate Committee Investigating Organized Crime and
the Illicit Traffic in Narcotics. Valachi was closely questioned by Chairman
McClellan on the very point of the Purge. When asked by the Chairman
how many men were killed, Valachi responded, "four or five, Senator."(8)
The Committee was obviously concerned about the number of men killed
that day, for almost immediately after Valachi's statement that four or five
were killed, the Chairman asked Police Sergeant Ralph Salerno to take the
stand. Salerno, a recognized expert on organized crime, was asked if he had
any information about other murders (besides Maranzano's and James Le
Pore's who was identified by Valachi as one of the four or five victims) that
took place that day. Salerno replied no. Wanting to be sure, the Committee
asked Salerno the same question once again. And again, Salerno answered
that he had no record of any other murders occurring that day.(9)

While all the believers in Purge Day cite figures of from thirty to
ninety men executed, and while all believers cite or refer to Valachi as
evidence, Valachi only testified about four or five murders. It should also be
noted that Valachi's recollections about the murders were exceptionally hazy,
undoubtedly because of the intervening time, but equally because his
knowledge was based on heresay. What has been done is to use the popular
story of Valachi as a primary source in the reconstruction of the history of
organized crime.(10)

There is another important and related question to be asked about the
Purge story. If Valachi did not testify about it, and if he is the major source

of the history of organized crime, then how did the story evolve? Certainly it seems obvious that the Senate Committee had heard of the story from the nature of their questions. There really is no mystery, however. The story of the Mafia Purge has been around since the late 1930s when it first appeared in a series of articles written by J. Richard "Dixie" Davis and published in Collier's magazine. In the third installment of his life and times, Davis, who had been Dutch Schultz's attorney, recounted a story told to him by Schultz mobster, "Bo" Weinberg. In an apparent moment of trust and confidentiality, according to Davis, Weinberg who was supposedly one of the killers of Maranzano confided that "at the very same hour when Maramanenza [sic] was knocked off . . . there was about ninety guineas knocked off all over the country. That was the time we Americanized the mobs."(11) Davis did add the caveat that he had been unable to check on the accuracy of Weinberg's claims about mass murder.

The unsubstantiated story of the Purge remained in limbo for approximately a decade. It resurfaced in one of the key books on organized crime published in 1951. Murder, Inc., written by ex-Assistant District Attorney (Brooklyn) Burton B. Turkus and Sid Feder, revitalized the story and transformed it into a major turning point in the history of organized crime. Turkus and Feder wrote: "The day Marrizano [sic] got it was the end of the line for the Greaser Crowd in the Italian Society . . . a definite windup to Mafia as an entity and a power in national crime." They added that "some thirty to forty leaders of Mafia's older group all over the United States were murdered that day and in the next forty-eight hours."(12) One of their primary sources for this part of their history was "Dixie" Davis's story of his conversation with Weinberg. Turkus and Feder also claimed that "even more irrefutable evidence was provided . . . by an eagle-beaked, one-eyed thug named, Ernest Rupolo.(13) A convicted murderer turned informer, Rupolo supposedly "went into the background and modernization of the Italian Society of crime." Turkus and Feder gave no details of the Rupolo story simply implying that it was consistent with the Davis account and their particular notions about the history of organized crime. Even if they had supplied particulars, however, the Rupolo history would have been no more credible than the Davis one. Rupolo, who was a rather unsuccessful informer, was not a participant in either the War or its supposed culmination, the Purge. In 1931 he was either fifteen or sixteen years old and whatever knowledge he may have had about the inner workings of the underworld was at the most based on rumor. In addition, Rupolo was something of a dope and a criminal incompetent. In 1934, he shot someone at the request of the gang. The victim survived and Rupolo went to jail serving nine years before realizing that the protection promised him by the gang was not forthcoming.(14)

There is one final personality or alleged source for the Purge story. In 1971, reporter and crime writer, Hank Messick, published a biography of Meyer Lansky, one of America's most famous criminals. In it Messick recounts the story of the Purge only this time its accuracy was based on the supposed confessions of a retired Mafia leader, Nicola Gentile. During the late 1920s and 1930s, according to Messick, Gentile was something of a Mafia peacemaker. Messick noted that Gentile's account of the gangland war and Purge was part of a long confession given to the FBI which he, Messick, was fortunate enough to examine. "Thanks to Gentile," Messick wrote, "there exists a firsthand account never before published," etc.(15) For those not so fortunate to have access to FBI files, however, the Gentile story can be read in an Italian paperback titled <u>Vita di Capomafia</u> which was published in the 1963. Gentile's story is a vast history of the American Mafia overwhelmingly stuffed with the names of almost every leading gangster, the dates and places of innumerable meetings and conferences, and the trivia of gangland diplomacy and violence. Gentile's devotion to detail is faithfully followed through his description of the murder of Maranzano. But then, as if in an afterthought, Gentile stated that the death of Maranzano was the signal for a massacre. But not a name, not a place, not a killing is described or given. As far as historical evidence is concerned the Gentile account provides no more evidence for the Purge than the earlier ones.(16)

It is one thing to question or dismiss the historical sources for the story of the Purge, it is something else to prove that it, in fact, did not happen. That is, while the sources may be incompetent, the event might still have taken place. In order to investigate that question, I did a survey of newspapers in selected cities beginning two weeks prior to Maranzano's death and ending two weeks after. I looked for any stories of gangland murders that could remotely be connected with the Maranzano case. The cities chosen were New York, Los Angeles, Philadelphia, Detroit, New Orleans, Boston, Buffalo and Newark. While I found various accounts of the Maranzano murder, I could locate only three other murders that might have been connected. Two of the killings were extensively reported in the Newark <u>Star Ledger</u>. The dead men were Louis Russo and Samuel Monaco, both New Jersey gangsters. Monaco and Russo were two of the men mentioned by Valachi in his testimony about the four or five murders. The other case that might be connected was found in the Philadelphia <u>Inquirer</u> on September 14. Datelined, Pittsburgh, September 13, an A.P. wire story told of the murder of Joseph Siragusa whose death was "attributed to racketeers" who "fled in an automobile bearing New York license plates."

The killing of four or five men does not make a purge, and certainly the killing of three or four men in the New York Metropolitan area and one man in Pittsburgh does not make a national vendetta. It is also significant that all the names turned up in our survey of the historiography of the Purge story--Maranzano, Monica, Russo, Le Pore, and perhaps Siragusa--have been accounted for. Left out are only the fictional members of the Mafia's Legion of the Damned--those unnamed and more importantly, unfound gang leaders whose massacre signaled the end of one criminal era and the beginning of another.

Earlier I asked the question that if Valachi did not testify about the Purge, and if he is the major source for this history of organized crime, then how did the story evolve. Having answered that question by the discussion of Davis, Turkus, and the others, however, there is still another puzzle left. If all the accounts of the Purge are unreliable, and if the newspaper survey is accurate, and finally, if the Purge never really took place, then how and why did various people concoct it? Did Weinberg simply want to spice up his account of the murder of Maranzano with tales of ninety other dead Italians? Did Rupolo simply want to further ingratiate himself with the Brooklyn District Attorney's office, and therefore invent a story? The same question can be posed for the other principals. I think the answer is no; I believe that they believed the story that they told.

All the commentators mention that the early 1930s was a time of intense confusion in the underworld compounded by the murders of such leaders as Joe Masseria in April, 1931, and Salvatore Maranzano in September, 1931. If not an actual war, there clearly was violence and division among Italian- and Sicilian-American gangsters. One can reasonably surmise that the death of Maranzano, especially to interested outsiders and/or principals such as Weinberg and Gentile, and lower-level, young hoodlums like Rupolo, was a momentous event that both could have and should have signaled increased violence. Certainly the level of anxiety along with the need for a comprehensible framework was high. When increased violence did not follow Maranzano's death, a comprehensible framework was established, not out of whole cloth, but out of the bits and pieces of events that were transformed as they were transmitted. The suggestion, therefore, is that the Purge story performed the function of reducing anxiety by magically wiping out Maranzano's followers--those who would have been expected to revenge their leader's death. Furthermore, there is evidence that suggests why criminals, especially in New York City, developed this particular story.

In the course of my newspaper search for Purge victims, I found a number of stories about Maranzano and his presumably major racket: the importation of aliens. In the September 10 edition of the Newark Star Ledger, for example, there was a story headlined, "U.S. hot on trail of ring busy smuggling aliens." Datelined, New York, the story noted that the Federal government was seeking indictments "against more than a score of persons in the hope of breaking up an alien smuggling ring . . . believed responsible for importing 8,000 foreigners in this country." Reportedly, nineteen men had already been indicted. "Headquarters were maintained," according to the paper, "in Montreal," while other administrative offices "were established in New York, Chicago and San Francisco." Other cities mentioned in the article included Detroit and Buffalo as transshipment points for the smugglers. On the following day, September 11, the Star Ledger headlined a story, "murder in N.Y. held reprisal by smugglers." The report noted that "authorities . . . were convinced that Salvatore Maranzano slashed and shot by three assassins . . . was the victim of the ring." The Star Ledger went on to say that the police were sure that members of the ring had murdered Maranzano for supposedly informing on them.

In the Los Angeles Times story of the Maranzano killing it was reported that detectives had "found immigration blank forms, and Department of Justice agents expressed the belief Maranzano had been engaged . . . in aiding aliens to enter the United States illegally." Finishing the Maranzano account, the L.A. Times wrote that "his slayers slashed a cross on his face--the sign of a traitor." Maranzano's link to alien smuggling was also reported by the New York Herald Tribune which stated on September 12, that "among the theories engaging the attention of investigators is one that Maranzano was marked for death in the belief that he had given evidence or was about to do so against his fellow conspirators in the smuggling racket." The Herald Tribune pointed out that his death "came a week after the arrest of nineteen men, . . . on charges of smuggling 8,000 aliens across the Canadian and Mexican borders, at a price ranging from $100 to $5,000 a head." And finally, The New York Times reported on the connection between Maranzano and alien smuggling. The New York Times story on September 11, noted that detectives were sent to Buffalo, Chicago, Poughkeepsie, New York, and Sea Isle, New Jersey, following Maranzano's murder. "The last named point," the Times added, was "said to have served as a base in the smuggling operations." On September 12, the Times also stated that Murray W. Garrson, Assistant Secretary of the Department of Labor, admitted that Maranzano had been under surveillance for over six months.(17)

In what way does his material on Maranzano contribute to the development of the Purge story? First, it is significant that this particular racket was national, or rather international, in scope. There are a number of reports that identify locales from at least Chicago to New York as both crossing points for aliens and headquarters of the racket. In addition, there are stories which indicate that law enforcement agents were sent to a number of different cities for purposes of investigating either the murder or Maranzano or the ramifications of this matter. It is possible, if not probable, that this kind of geographical activity accounted for the notion that the Purge was national. Surely, something was happening all over the country because of Maranzano's death. Secondly, the murder of Maranzano, if indeed connected to the investigation of the alien smuggling racket, could have caused the enterprise to cease, at least for an important period of time. If so, his death could be interpreted as resulting in the end of the traditional Mafia which, if it existed, must have depended on the continuous importation of Sicilians. Thirdly, it can easily be imagined how the deaths of four or five supposedly important racketeers could be transmuted into a massacre of large proportions through the mechanisms of rumor and hyperbole, standard fare in the secretive oral culture of the underworld.

It is possible, therefore, to explain the belief in the Purge on the part of various underworld figures. It is also clear how the story developed and that Valachi was, in a sense, forced to bear witness to a story of which he was barely cognizant. But, it is also by no means clear why so many scholars have bought a story which so obviously   violates historical respectability. Without speculating for the moment on the devotion to "history as conspiracy," it is surely important to note that the scholarly attachment to the Purge is a conspicuous example of the insensitivity to historical methods found all too often in work on organized crime. Unfortunately, it is not enough to point out examples of historical naivete and insensitivity with the admonishment for care and attention to historical methods. Much of the contemporary sociology of organized crime has been constructed from the interpretive framework of the popular histories as well as from their narratives. Academic sociology has displayed a strong affinity with the "ideological preconceptions" of the creators of the popular works even when it shifts gears by focusing on different criminal personalities as the architects of organized crime. The best example of this tie is found in the elegant and highly influential work of Donald Cressey which begins with the Purge and is seemingly held together by a rigorous logic which masks the shared preconceptions.

Cressey stated: "To use an analogy with legitimate business, in 1931 organized crime units across the United States formed into monopolistic corporations, and these corporations, in turn, linked themselves together in a monopolistic cartel." Along with corporate organization went political confederation: "To use a political analogy, in 1931 the local units formed into feudal governments, and the rulers of these governments linked themselves together into a nation-wide confederation which itself constitutes a government."(18) Cressey's sociology quickly moved past 1931 and history, to its major analytical engine. As both business and polity the single most important social fact of organized crime is its "division of labor" which is the structure of organized crime. This division of labor is the cornerstone of Cressey's thinking. He wrote that "an organized crime is any crime committed by a person occupying a position in an established division of labor designed for the commission of crime;" and "the organized criminal, by definition, occupies a position in a social system, an 'organization,' which has been rationally designed to maximize profits by performing illegal services and providing legally forbidden products demanded by the members of the broader society in which he lives."(19)

From the perceived division of labor, Cressey selected a particular position, the Enforcer, for analysis. The occupants of this position, he wrote, arrange "for the injuring or killing of members" of organized crime and outsiders. Once found, discussion of this position "enables me to create information about complex governmental processes and a set of laws." The Enforcer is "one of a subset of positions existing within a broader division of labor" whose function is "to maximize organizational integration by means of just infliction of punishment of wrongdoers."(20) Moving from the area of structure to governance, he stated: "The presence of an Enforcer position . . . also can be taken as evidence that members of the organization must have created some functional equivalent of the criminal law, . . ." Further inferences drawn include the "fact that punishments are to be imposed 'justly,' in a disinterested manner." Building on this inference, it is then held that "the relationships between organized criminals are, to a great extent, determined by rules and expectations which insure that when "justice prevails, the norms that govern the resort to adjudication serve to reduce conflict . . . ."(21)

Where all of this is heading, of course, is toward the final statement on the structure of organized crime. The full-blown claim is that the structure is characterized by a more-or-less "totalitarian organization" with "rigid discipline in a hierarchy of ranks" with "permanency and form" which extend beyond the lives of particular individuals and "exist independently of any particular incumbent."(22) Cressey stands as the most forceful exponent of the view that organized crime is a bureaucracy and that syndicate criminals were at least protobureaucrats. This criticism of Cressey's analyses of organized crime is primarily concerned with pointing out the inevitable mistakes of those who rely on the conspiratorial biases of the popular interpreters.(23)

With just a few changes, Cressey's seemingly inferential sociology could have been lifted from the pages of Murder, Inc. All the categories enumerated in it, including the fixation on Enforcers as the key to the nature of The Organization, can be found in Chapter Four, "National Crime: A Cartel.": And just as the Purge story related by Turkus and Feder was provided by an informer, so their elaborate sociology was constructed from the testimony and recollections of another informer, Abe "Kid Twist" Reles. But unlike Rupolo and the other sources for the Purge story, Reles was deeply involved in most of the events he related and which, Turkus and Feder hold, established the single national ruling cartel in organized crime. Reles was an extraordinarily powerful witness sending four men to the electric chair, but what he testified about and endlessly discussed while held as a material witness were particular events. What Turkus and Feder wrote about almost a decade later was their vision of the ordering of those events, a vision that was as neat and tidy as the contemporary sociologies of Mafia/Cosa Nostra. The framework brought to bear on Reles's confessions was the lawman's favorite: the ladder of conspiracy with each rung integrated in a series leading to the Master Conspirator.

The problem with the sociology of organized crime advanced in Murder, Inc.. is that it was wrong, as a long and careful search through all the trial transcripts and extant internal documents from the investigation of murderers in Brooklyn reveals.(24) What Turkus and Feder described and concluded was simply not what Reles related. For instance, when Reles talked about cooperation in the Brooklyn underworld, Turkus and Feder called it organization. When Reles talked about favors being done by one racketeer for another, Turkus and Feder wrote about orders and a smooth chain of command. When Reles talked about the geographical mobility of various criminals, Turkus and Feder held it was proof of the national scope of the cartel. When Reles talked about the innumerable mobs that populated the New York underworld, Turkus and Feder interpreted it as proof of the confederation of organized crime. When Reles talked about murder using the underworld argot of contract, Turkus and Feder concluded that murder was a real business conducted in the interests of the National Crime Syndicate and carried out by a special group of enforcers. And finally, when Reles talked about the shifting, changing, bickering, competitive, murderous social world of organized crime, Turkus and Feder surmised that this untidiness was only the inevitable and necessary fallout of the consolidation of organized crime. Everything, no matter how counter-factual, led to the big conspiracy, The Organization.

This credulous sociology, innocent of such notions as informal organizations and patron-client networks, fixed the sociological frame of organized crime around conspiracy as surely as its narrative provided a history. The historical insensitivity and sociological primitivism of the popular account have been carried forward and incorporated into the realm of academic scholarship where they have inspired further mistakes. For example, Mark H. Furstenberg in his 1969 article, "Violence and Organized Crime," which was part of the Staff Report presented to the National Commission on the Causes and Prevention of Violence displays all the misguided tendencies previously discussed. Furstenberg began his analysis noting that "it is true that during its period of rapid change and development, when it was consolidating its strength, organized crime was violent, sometimes wildly so."(25) Presumably as evidence for this assertion, Furstenberg had started his study with the statement that "in Chicago, between 1919 and 1934, there were 765 'gang murders,' an average of 48 per year. From 1935 to 1967, a period twice as long, there were 229 murders, an average of seven per year."(26) Furstenberg's figures are meaningless, no conclusions are warranted. Unanswered in his work are the following questions: what are the figures year by year; what was the trend from 1919 to 1935; was murder fairly constant until 1935 and then did it decline or was there a steady decline, and so on. These important questions have been tentatively explored, I might add, by other scholars such as Henry Willbach who reported for New York City a fairly steady drop in the rate of homicides from World War I until the mid-1930s.(27)

Turning from this topic, Furstenberg went on to say that "being far more inherently rational than they are inherently violent, organized crime's leaders, . . . developed and refined a system of rational alternatives to violence.(28) There is virtually no evidence cited to support this conclusion.

In fact, shortly after this statement in a footnote, Furstenberg remarked that the "rationality of this decision can be appreciated only by realizing how inherently violent syndicate criminals really area.   They come from a background in which violence is used naturally and easily to settle disputes.(29) Not only is it unclear from the text and the footnote whether syndicate criminals are far more inherently violent or rational, but there is no discussion of their backgrounds or any documentation for either claim. There is no doubt, however, what ethnic group Furstenberg is alluding to-- Italian-Americans.

His next section contains a not so subtle shift in language in which organized crime becomes transformed to Mafia and then to La Cosa Nostra. As in most institutions, Furstenberg wrote, "organized crime's evolution from 1920 to 1940 was a mixture of conscious decision and unconscious response to changes occurring rapidly in most American institutions. The old Mafia was suited to a rural society; but the United States was becoming urban." After commenting on the increasing sophistication and organization on the part of the American Mafia, Furstenberg added: "La Cosa Nostra, the dominant force in organized crime today, is a government, with all the forms of government," etc.(30) But while cataloguing the achievements of La Cosa Nostra in controlling violence, he noted that "before 1962, the Boston activities of the New England Cosa Nostra branch were inhibited by three strong, independent Irish criminal organizations," especially in the areas of loan-sharking and gambling.(31)   So much one might think for the interchangeability of the terms organized crime and La Cosa Nostra.

Furstenberg also discussed ethnic succession in organized crime. He stated that "it has always been the tough, ambitious, first generation criminals who have had the stomach for street-level operations. There are few first-generation Italians to populate the lower levels of organized crime." And he noted that "just as the Jews moved out to make room for the tougher and hungrier Italians, the Italians were now moving out for Blacks and others."(32)   The citation for this statement is Mark H. Haller's article, "Urban Crime and Criminal Justice: The Chicago Case." In the footnote, Furstenberg repeated Haller's breakdown of organized crime in Chicago in 1930 which showed that of the "108 top crime figures . . . 30 percent were Italian, 29 percent were Irish, 20 percent were Jewish, and 12 percent were Negro."(33)   Furstenberg then added in the footnote, with no supporting evidence, that "we know that 1930 was a turning point in American organized crime." But, there was no proof offered that confirmed that last statement and no evidence given for the contention made in the text about ethnic succession.

Furstenberg did not just make up the statements on this particular issue, nor on the other historical and sociological conclusions offered in his paper. He was either summarizing the sense of, or quoting directly from Murder, Inc., The Valachi Papers and the work of Donald Cressey. In the study of organized crime in any of its aspects, one "is not free to take his history or leave it alone. Interpretations of the present require a host of assumptions about the past. And the real choice, therefore, is between explicit history, based on a careful examination of the sources, and implicit history, rooted in ideological preconceptions and uncritical acceptance of local mythology."(34)   Reliance on unsubstantiated accounts and the lawman's ideological preconceptions has mired the study of organized crime in the bog of conspiracy, allowing the term itself to be carelessly transformed to stand for the monolithic organization of criminals.

This is not to suggest that all contemporary scholars have fallen into propagating the "Mafia myth" and its corollaries. There are some notable exceptions.(35)   But I do claim that the connection between the terms organized crime and alien conspiracy (complete with the Purge) runs so deep that employment of it increasingly implies acceptance of the conspiracy. To write about organized crime is to saddle oneself with at least the outline of the ineluctable drive toward consolidation and confederation. In fact, this problem has recently moved several historical researchers to abandon the term organized crime altogether in favor of the term illegal enterprises in their efforts to construct a more responsible history. One example is Mark Haller who notes that "at the heart of what is often meant by organized crime are types of enterprises that sell illegal goods and services to customers: gambling, prostitution, bootlegging, narcotics and loansharking." Researchers, Haller advises, should first "identify a type of criminal activity" and then explore the questions of organization and coordination in a variety of historical settings.(36) Haller's call, then, is for work that will build the historical foundation without which much contemporary theorizing will remain in the conspiracy trap.

* * * * * * * * * * * * * * * * * * * *

NOTES

1. Warner's methodology, as Thernstrom makes clear, was consciously ahistorical. This puts Warner in a decidedly different intellectual stance than our examples. Stephan Thernstrom, "'Yankee City' Revisited: The Perils of Historical Naivete," American Sociological Review (April 1965).

2. Thernstrom, 237-238.

3. David L. Chandler, Brothers in Blood: The Rise of the Criminal Brotherhoods (E. P. Dutton, 1975).

4. Donald R. Cressey, "The Functions and Structures of Criminal Syndicates," in Task Force Report: Organized Crime, President's Commission on Law

Enforcement and Administration of Justice (Government Printing Office, 1967), Appendix A, 26.

5. Chandler, 160.

6. Donald R. Cressey, <u>Theft of the Nation: The Structure of Crime in America</u> (Harper & Row, 1969), 44.

7. Cressey, <u>Theft of the Nation</u>, 36-37.

8. U.S. Senate, Committee on Government Operations, Permanent Subcommittee on Investigations, <u>Organized Crime and Illicit Traffic in Narcotics: Hearings</u> (Government Printing Office, 1963), 232.

9. Senate, 333.

10. Peter Maas, <u>The Valachi Papers</u> (Bantam Books, 1968).

11. J. Richard "Dixie" Davis, "Things I Couldn't Tell Till Now," <u>Collier's Magazine</u> (August 5, 1939), 44.

12. Burton Turkus and Sid Feder, <u>Murder, Inc.: The Story of the Syndicate</u> (Farrar, Straus and Young, 1951), 87.

13. Turkus and Feder, 88.

14. The information on Rupolo comes from <u>The New York Times</u>, June 29, 1944; August 16, 1944; May 6, 1946; June 7, 1946; and June 10, 1946.

15. Hank Messick, <u>Lansky</u> (G. P. Putnam's Sons, 1971), 49.

16. Nicolo Gentile, <u>Vita di Capomafia</u> (Editori ruiniti, 1963).

17. I have attempted to get the files on Maranzano from the Immigration Service. Unfortunately, they have been missing since late September, 1931.

18. Cressey, "Functions and Structures," 31.

19. Cressey, 29.

20. Donald R. Cressey, "Methodological Problems in the Study of Organized Crime as a Social Problem," <u>The Annals</u> (November 1969), 110.

21. Cressey, "Methodological Problems," 111-12.

22. Cressey, "The Functions and Structures," 58.

23. Joseph Albini in <u>The American Mafia: Genesis of a Legend</u> (Appleton-Century-Crofts, 1971) argues a strong case against Cressey and other holders of the notion of a bureaucratic structure for organized crime. Albini's critique is based on his finding of an anti Italian-American bias along with the use of limited and suspect sources.

24. The following comprise the <u>Murder, Inc.</u> material:

MANUSCRIPT COLLECTIONS

Dewey, Thomas E. Personal Papers. Department of Rare Books, Manuscripts and Archives, The University of Rochester Library. (Series 1: Early Career, Boxes 10 and 90 contain some material relating to the criminals who became known as the Murder, Inc., Enforcers).

New York State Crime Commission. "Papers Relating to the Commission's Investigation of Waterfront Crime, 1952-1953." Rare Book and Manuscript Library, Columbia University Library.

O'Dwyer, William. Personal Papers. New York City Municipal Archives. Included in the O'Dwyer papers are all the existing documents concerning the Murder, Inc., investigation. Among the <u>un-indexed</u> papers are summaries of Reles's testimony and recollections; memoranda of police investigations; grand jury minutes; selected parts of trial transcripts; arrest records of the different criminals held and tried; and various documents showing how Reles's story was fashioned to the preconceptions of the District Attorney's office.

TRIAL DOCUMENTS

County of New York, Court of General Sessions, Part V. The People of the State of New York Against Louis Buchalter, Max Silverman, Harold Silverman, Samuel Schorr. New York, January 26, 1940. Vols. 1,2,3,4. The Court's Copy 5709.

Brooklyn, New York, Court of Appeals. The People of the State of New York Against Louis Buchalter, Emmanuel Weiss, Louis Capone. May-June, 1942. 7944. Vols. 1,2,3,4,5.

Brooklyn, New York, Court of Appeals. The People of the State of New York Against Irving Nitzberg. September-October, 1941. 7760.

People v. Buchalter, 289 NY 181; 289 NY 244; 44 NYS 2d 449; 29 NYS 2d 621.
People v. Goldstein, 285 NY 376.
People v. Maione, 284 NY 423.
People v. Nitzberg, 287 NY 183.
United States v. Buchalter, 88 F. 2d 625.
United States v. Buchalter, FRC 539652, Box 332479.

REPORTS OF INVESTIGATIONS AND COMMITTEES
Amen, John Harlan, Report of Kings County Investigation, 1938-1942 (New York, 1942).

Citizen's Committee on the Control of Crime in New York, Inc. Crime in New York City in 1939 (New York: The Committee, 1939).
_____, How Rackets Work and How They May be Stopped, Together with the Text of an Address by Thomas E. Dewey, May 11, 1937 (New York: The Committee, n.d.).

New York State, County of Kings, Supreme Court, A Presentment Concerning the Enforcement by the Police Department of the City of New York of the Laws Against Gambling (Hamilton Press, 1942).
_____. A Presentment on the Execution of Bail Bonds (Hamilton Press, 1942).

U.S. Department of Justice, Federal Bureau of Investigation. "The Fur Dress Case," I.C. #60-1501 (Washington, D.C., November 7, 1939).

U.S. Department of Treasury, Bureau of Narcotics, Traffic in Opium and Other Dangerous Drugs (Government Printing Office, 1937-1941).

25. Mark H. Furstenberg, "Violence and Organized Crime," in Crimes and Violence: Staff Report, the National Commission on the Causes and Prevention of Violence, vol. 13, appendix 18 (Government Printing Office, 1969), 912.

26. Furstenberg, 911.

27. H. Willbach, "The Trend of Crime in New York City," Journal of Criminal Law and Criminology (1938).

28. Furstenberg, 912.

29. Furstenberg, 932.

30. Furstenberg, 913.

31. Furstenberg, 920.

32. Furstenberg, 912.

33. Furstenberg, 936.

34. Thernstrom, 242.

35. Among the best work is that of Francis A. J. Ianni, <u>A Family Business:</u> <u>Kinship and Social Control in Organized Crime</u> (Russell Sage, 1972); Dwight Smith, Jr., "Mafia: The Prototypical Alien Conspiracy," <u>The Annals</u> (January 1976); William J. Chambliss, "Vice, Corruption, Bureaucracy and Power," <u>Whose Law, What Order?: A Conflict Approach to Criminology</u> (John Wiley, 1976); Mark H. Haller, "Organized Crime in Urban Society," <u>Journal</u> <u>of Social History</u> (winter 1971-1972); William H. Moore, <u>The Kefauver</u> <u>Committee and the Politics of Crime, 1950-1952</u> (University of Missouri Press, 1974); and Anton Blok, <u>The Mafia of a Sicilian Village, 1860-1960: A</u> <u>Study of Violent Peasant Entrepreneurs</u> (Harper & Row, 1974).

36. Haller's remarks are contained in the draft of a paper, "The Rise of Criminal Syndicates" which he generously sent in the spring of 1975.

Chapter 2
Contemporary Reports on Organized Crime:
Errors of Historical Interpretation

This essay examines certain historical assumptions found in statements by federal prosecutors, other officials in law enforcement, and government protected informants on the subject of organized crime. Many of the current assumptions have arisen from the ever wider-ranging employment of Title IX of the Organized Crime Control Act passed in 1970 (and amended since then) commonly known as RICO standing for Racketeer Influenced and Corrupt Organizations.(1) RICO is in one sense a conspiracy statute and its use compels the government to establish particular, discrete conspiracies. Thus prosecutors are searching for patterns of linked criminal activities; not linked at a particular moment in time but in meaningful ways over time. Duration is quite important and that is fine in those cases and statements where it is appropriately supported by reasonable evidence. But conspiracy thinking is often too expansive, casting a wider historical net than data can support. When prosecutors and their colleagues (including some academic researchers), whether in case presentation or in public statements, talk about "classic" Italian/American organized crime their sense of history and conspiracy often is excessive and extreme.(2)

HISTORY AND LA COSA NOSTRA

Federal prosecutions of organized crime in the 1980s seemed to many to have nailed down forever the issue of the history and structure of "classic" organized crime.(3) For instance, in the civil RICO case pressed by the U.S. Attorney in New York's Southern District against the Teamsters' Union in an attempt to break "La Cosa Nostra's control over one of the largest labor unions in the nation," Assistant U.S. Attorney Randy M. Mastro stated the following: "The Government has established the existence of La Cosa Nostra, a/k/a the 'Mafia' or 'Our Thing'" partially "through the testimony of Angelo Lonardo, former Underboss of the Cleveland Organized Crime Family."(4) Mastro remarked that Lonardo established the subsequent items concerning La Cosa Nostra: (1)it operates throughout the U.S.; (2)it has a Commission which is made-up of the bosses of New York's five La Cosa Nostra Families and "at various times, representatives of other La Cosa Nostra Families throughout the United States"; (3)the Commission is a "national ruling council" of Mafia groups; (4)the Commission's function includes regulating, facilitating and controlling relationships among Mafia groups.(5)

Historians' interest in the above concern whether a fair reading of Lonardo's statements supports the contentions. If not, they suffer in this instance from what historian David Hackett Fischer called "the fallacy of the pseudo proof."(6) This is not to doubt what Lonardo said at various trials, nor to call into question the convictions on many substantive crimes of the organized criminals tried. It does question whether his testimony on particular historical issues jibes with what he said to the FBI before the trials. Historians must seek the best relevant evidence possible and in Lonardo's case that would be his recollections prior to the undoubted coaching preparatory to his many trial appearances as a government witness.

Lonardo was interviewed about the "history of the Cleveland La Cosa Nostra (LCN) family" by Special Agents Richard B. Hoke and R. Gerald Personen on 28 October 1983.(7) At the very beginning of his story he noted his Mafia boss through the mid-1970s, John Scalish, "was not familiar with the names and activities of his peers in other families around the country." When Scalish travelled he would have to ask Lonardo for factual information about other bosses. Scalish died in May 1976 and had to be replaced. The successor was James Licavoli who at first demurred claiming he wasn't nearly as qualified as Lonardo was for the position. In fact, Lonardo stated to his FBI inquisitors that "Licavoli was completely ignorant of certain matters of protocol required by his new position and it was he (Lonardo) who advised Licavoli that he had to go to New York to advise the boss of the Genovese family of his new position and to pay his proper respects."

Later on Lonardo was asked about his own knowledge of "La Cosa Nostra (LCN) families in other cities." He answered that New York had five and there was one in each of the following American and Canadian cities: "Los Angeles, San Francisco, Denver, Milwaukee, Chicago, Kansas City, Detroit, Cleveland, St. Louis, Buffalo, Philadelphia, Boston, Providence, Pittsburgh, New Orleans, New Jersey [sic], Windsor, Ontario, Toronto, Ontario, Montreal, Quebec." Lonardo added that the Chicago Mafia was "responsible for settling certain disputes and making administrative type decisions that affect the various families in the Western half of the United States." The Eastern region was the responsibility of New York's Genovese Family with certain important exceptions. Eastern waterfront organized crime, for example, was controlled by New York's Gambino Family.

With the Chicago mob administering western Mafias and the Genovese gang doing much the same for eastern ones, it is a fair question to wonder precisely what the Commission was empowered to do. This must have been on the interviewing agents' minds also for they next wrote "Lonardo acknowledged the existence of a national commission of LCN bosses which meets on rare occasions to settle disputes or make high-level decisions involving national LCN matters" (my emphases). And although Lonardo was far and away the most knowledgeable Cleveland Mafioso when it came to national LCN matters it turned out he only knew two members of the Commission. They were both from New York.

These statements cannot stand as confirmation of any of the general principles enunciated by the government about La Cosa Nostra in its case against the Teamsters.   Indeed, when questioned about the Teamsters' involvement with LCN Lonardo had this to say about his boss Licavoli's knowledge: "Licavoli asked Lonardo why the Cleveland LCN family cared who became head of the Teamsters since they were not getting any money from the Teamsters." Amusingly, Lonardo and an associate Milton Rockman couldn't find on their own a Chicago office where they were supposed to discuss Teamster politics with the Chicago LCN. They had to ask directions from someone at a dry cleaning shop who warned them "they were in a bad neighborhood and would be lucky to get out alive." No matter what Lonardo said about these matters later, it seems that in October 1983 he wasn't very much more aware of the LCN Commission than his two bosses. What made his memory improve later is anyone's guess.

This brief excursion into the omnipresent claim of a slick domineering Commission serves to introduce wider historical themes. Statements about La Cosa Nostra and the Commission at the Commission trial in New York, in testimony before President Reagan's Commission on Organized Crime, and later in hearings held by the Senate's Permanent Subcommittee on Investigations reveal serious historical (and conceptual) errors. There are several key dates, for example, which dominate the issue. The earliest was given by the Justice Department's Organized Crime Intelligence and Analysis Unit which operates within the Criminal Investigative Division's Organized Crime Section. The Unit presented to the Permanent Subcommittee on Investigations a "Chronological History of La Cosa Nostra in the United States" which began with the 1890 murder of David Hennessey who was Superintendent of Police in New Orleans. This murder and subsequent ones in New Orleans stimulated by it "created perhaps the first significant public awareness of the La Cosa Nostra (LCN)."[8] It is well to note at the start of this discussion that this statement is likely an example of the "fallacy of ambiguity which consists in the use of a word or an expression which has two or more possible meanings, without sufficient specification of which meaning is intended," mixed with the error of "prolepsis" which "describes an event as happening before it could have done so."[9] It is simply not true that the Hennessey murder made anyone aware of La Cosa Nostra as a term signifying organized crime for it was probably not invented and certainly not popularized for another 73 years. Even the Intelligence and Analysis Unit knew the statement was factually incorrect for it followed this preposterous claim noting this event triggered the introduction into American society of the term Mafia.

One may wish to argue at some point such as the Commission trial in the 1980s that Mafia and La Cosa Nostra are synonymous, but one cannot argue that until both terms are invented. Of course the Intelligence and Analysis Unit was merely looking for some starting point to present its "chronological history" of LCN and I presume no one was to take the Hennessey murder as actually signifying anything very serious.(10) The Intelligence and Analysis Unit itself had a later date in mind which was really significant for the formation of La Cosa Nostra. I should note that this penchant for reaching as far back into the past as possible to establish the first instance of the criminal conspiracy later called La Cosa Nostra was also an intriguing feature of the "Commission" indictment which charged that "from 1900 to the present there existed a nationwide criminal society known as La Cosa Nostra or the Mafia," etc.(11) Needless to add, the criticism of the Intelligence and Analysis Unit's historical sense holds for the indictment.

LCN's important history, commented the Unit, started 30 years later with Prohibition "which enabled the small, but powerful LCN to capitalize upon its international contacts, its reputation for ruthlessness, and--above all--its rigidly disciplined structure of cooperating gangs to establish the position of unrivaled eminence it holds in the American underworld today."(12) To prove this mini-thesis, the Unit's history then rattles off murder after murder beginning with the killing of Big Jim Colosimo on 11 May 1920 and ending with the April 1931 shooting of "Giuseppe Masseria, "boss of all bosses." But this decade of reported mayhem cannot sustain that part of the thesis which holds the LCN was by 1920 small but composed of rigidly disciplined cooperating gangs. If anything, the facts marshalled by the Unit indicate the opposite. In fact, they are so confused and antithetical to their task as to mean either the opposite or nothing at all. New terms are thrown in without explanation such as "Camorra" which is said to have merged with LCN in Chicago on 7 September 1928 following yet another killing.(13) No evidence is developed through this list of violence that the LCN had international contacts, another important part of its alleged development.

There are other interpretative errors accompanied by bizarre "non sequiturs" which need comment such as the following datelined 17 June 1933. "FBI Agent, Raymond J. Caffrey, 3 police officers, and hoodlum Frank Nash were killed in the Union Station parking lot during [sic] the infamous 'Kansas City Massacre.' Reportedly the Kansas City LCN Family declined a request that it participate in the operation."(14) This factoid cannot establish anything about the LCN even if it was true (which is not actually claimed) that someone asked it to participate in several murders. It is a variant of the "prodigious fallacy" which "mistakes sensation for significance" and an example of "negative evidence" which is really no evidence at all as David Hackett Fischer has noted.(15) To the best of its knowledge, the Unit had nothing positive to say connecting the LCN to the "Kansas City Massacre." It wasn't involved, it didn't participate, thus what could the event possibly tell about LCN? The answer is nothing.

By far the most important error, however, lies in the Unit's resurrection of perhaps the best known false claim in historical studies of organized crime. This is the alleged transformation of LCN structure following the murders of Masseria and on 10 September 1931 of Salvatore Maranzano.(16) It is the most infamous example of the discredited "alien conspiracy myth" as Dwight Smith, Jr. called it.(17) A tale of murder and the re-organization of LCN in 1931, it never had credible sources as several writers pointed out years ago.(18)

The burden of establishing historical accuracy for the development of "classic" organized crime exceeded the grasp of the Intelligence and Analysis Unit whose work has been under consideration. (The same holds true for the analysts in the Intelligence Division of the New York City Police Department who presented the same story in attenuated form to the Permanent Subcommittee on Investigations.)(19) The Unit likely figured a "potpourri" of mob facts placed in chronological order would somehow reveal the actual essence of organized crime--La Cosa Nostra. In this quest the writers thus also commit the "fallacy of essences" which emerges from the notion that everything (in this case organized crime) contains deep inside an essence, "an inner core of reality" (in this case La Cosa Nostra).(20) For those who proscribe to this the facts about a culture, an ideology or an institution (even a secret criminal order) are only important to the degree they demonstrate the essence of the subject. But this is an essentially fruitless endeavor as philosopher Karl Popper remarked in noting that the "progress of empirical knowledge requires, not a search for essences, which cannot be found by any empirical method, but rather a search for patterns of external behavior."(21)

THE SICILIAN ANGLE

In the 1988 Permanent Subcommittee Hearings provocatively titled Organized Crime: 25 Years After Valachi, a fairly recent Sicilian Mafia defector, Tomasso Buscetta, appeared as a principal government witness. In introducing him it was pointed out that Buscetta's 1986 testimony in Sicily "was instrumental in the conviction of 435 members of the Sicilian Mafia," and the year before that his testimony in the "famous Pizza Connection case in New York City helped to convict 35 members of the New York and Sicilian La Cosa Nostra."(22) He was appearing before the Senate Panel to relate what he knew about the history of the Sicilian Mafia/La Cosa Nostra and in particular its relations with its American counterpart. Buscetta, who came from Palermo, first stated his credentials:

> I am a member of the Mafia, or Cosa Nostra, in
> Sicily, known within the Cosa Nostra as a Man of
> Honor. Since 1948, I have been part of this
> organization both in Sicily and later as a favored

and protected guest in the United States. I have
known among others Salvatore, or "Lucky"
Luciano, Carlo and Paolo Gambino, Paul
Castellano, and Joe Bonanno. I have spent many
years in prison for my activities as a Mafioso. I
have lost one brother, two sons, one brother-in-
law, three nephews, and one son-in-law all at the
hands of the Mafia.(23)

There are two historical issues I wish to consider in the discussion of
Buscetta. The first has to do with his account of the formation of a Sicilian
Mafia/La Cosa Nostra Commission, and the second with claims about more
fundamental historical changes within the Sicilian Cosa Nostra itself.

On the first issue Buscetta's statement is quite clear. The inspiration
for a Sicilian Commission came from none other than Lucky Luciano.
Buscetta holds that following Luciano's deportation from the U.S. to Italy, "in
the early 1950s," he and Luciano became fast friends. During this period
Luciano supposedly told him "how and why he created the commission in
America." The reason was to prevent the "killing of Men of Honor unjustly."
With this in mind the Commission was formed and "had the bosses of all the
families in the United States as its membership," although this would have
been news as we already know to Cleveland's "Men of Honor." As it was
formed to stop indiscriminate killings, the American Commission's essential
function, Buscetta explained, was the organization of sanctioned mob
murders.

It is difficult to take this seriously for several reasons. As everyone
interested in Luciano knows, except apparently his pal Buscetta, Luciano was
deported from the U.S. early in the winter of 1946.(24) Moreover, he spent
the greatest amount of his Italian exile in Naples not Palermo although this
wouldn't preclude his chatting with Buscetta when in Palermo. Nevertheless,
not knowing the date, indeed the era, of Luciano's deportation raises concern
about the historical credibility of Buscetta's testimony.

In addition, the hoary claim that rationalizing murder was the first
cause in Commission formation still lacks convincing proof. While this could
have been on Luciano's mind, especially as a method of protection for he
had been involved in several bold murders in 1931, there is no credible
evidence that this possible desire led to real organizational innovation.(25)
The above-mentioned "Chronology" presented by the Intelligence and
Analysis Unit lists murder after murder following the time when Luciano
would have had to form the Commission (before he was arrested, tried,
convicted and sent to prison in 1935 and 1936). Although a few of the many
listed killings were allegedly carried out after some form of consultation
among gangsters, most appear not to have been, and there is no evidence
presented that this was so. Indeed, there aren't any adequate discussions of

the Commission's decision-making process which could lead one to this conclusion.

What I would offer, instead, is that powerful gangsters are murdered by competitors who often conspire with other powerful gangsters to try and insure their safety after the fact. Less powerful ones are killed more often, but then there are so many more of them. Murdering the weaker ones is usually not a problem unless they are money-makers for other gangsters who sometimes seek revenge because of relatively empty pockets. Bosses quite rightly are insecure about the loyalty of their own underlings wondering in times of trouble who is loyal and who is a traitor. And as the historical record of violence shows, they have good reasons to wonder. The notion that a national Commission was formed to rationalize murder (or more realistically to protect crime bosses), whether someone actually believed this was a good idea or not, is empirically unsupported. Moreover, even if this was part of the original motivation and could be established as such, the obvious fact that bosses have not been protected nor organized crime murder rationalized (whatever that could mean in such a treacherous social world) should have stopped this claim long ago. That it still persists, as Buscetta's testimony makes clear, indicates it serves other purposes or represents an "idealization" of organized crime. There is, of course, a third possibility: an "idealized" version of organized crime's development may well serve the bureaucratic needs of state agencies and fit, as well, with popular but erroneous social science paradigms; in this case the paradigm of modernization.(26) This, it seems to me, is most likely; the state and social science reinforcing each others' needs and prejudices.

The fact is this theme--sanctioned murder leading to a more complex criminal organization or the other way round--gained a foothold among criminologists because it seemed so logical, so ineluctable. Its attractiveness was such it wasn't noticed that it willynilly reversed the link between data and interpretation. Underworld tattlers filled in the appropriate empirical slots in the already extant theoretical structure.

THE MEANING OF VIOLENCE

One of the most cogent criminological statement that reveals this backward development was actually given several decades ago by Robert T. Anderson. In "From Mafia to Cosa Nostra," published in the American Journal of Sociology, Anderson argued that the Sicilian Mafia was changing as Sicily itself underwent the process of modernization. Part of that process can be seen, Anderson remarked, in the Mafia's adaptation and expansion of its "techniques of exploitation."(27) Quoting from a 1960 article by reporter Claire Sterling, Anderson found that the Mafia had become urbanized: "'Today there is not only a Mafia of the feudo (agriculture) but also Mafias of truck gardens, wholesale fruit and vegetable markets, water supply, meat, fishing fleets, flowers, funerals, beer, carrozze (hacks), garages, and construction. Indeed, there is hardly a businessman in western Sicily who doesn't pay for the Mafia's 'protection' in the form of 'u pizzu'."(28) It is not certain what is urban about all these activities (fishing fleets, for instance), but it is plain that Mafia enterprises had widened from what earlier accounts had suggested.

When all was said and done, the dramatic point was the Mafia "has bureaucratized," or almost bureaucratized, or was on the road to bureaucratization. Anderson wrote the old or traditional Mafia was "family-like," lacking at least three of the four necessary "characteristics of a bureaucracy."   However, that was all changing as Sicily "poised for industrialization with its concomitant changes." But because Sicily was only poised on the verge of modernization, the Sicilian Mafia was momentarily, at least, a mix of the old and the new. There was no doubt, though, where it was heading.

The American Mafia provided the example and model. It had achieved bureaucratization, remarked Anderson, "beyond that even of bureaucratized Sicilian groups." One of the key elements in this process involved a striking change in Mafia custom: "Modern mafiosi avoid the use of force as much as possible, and thus differ strikingly from old Sicilian practice."(29) Here, Anderson intimates that the contemporary Sicilian Mafia is also moving in the same direction--away from violence toward other methods of dispute settlement. His discussion of the American Cosa Nostra holds that this overarching structure which is both a proof and attribute of Mafia bureaucratization serves to "adjudicate disputes . . . to minimize internecine strife, rather than to administer co-operative undertakings."

The meaning of La Cosa Nostra, the function of the Commission, the movement of history, was to minimize violence, to rationalize murder, to routinize homicide. No matter how it was said or who said it--the direction of criminal organization was tied to an interpretation of murder. And because the belief in La Cosa Nostra was so tenaciously developed and held, the characteristics of murders committed by organized crime simply had to fit the theory. Data was captured by theory. Depending on who was speaking for these developments, certain twists were added. But they were of no consequence until someone looked at organized crime murders to see whether they fit the claims made. With one exception in Sicily, no one of any consequence has.(30) Thus when Mafia wars erupted and bodies seemed to be falling everywhere, there was no change in the theoretical structure built upon a serious historical misunderstanding.

But theory is, if anything, flexible; and it seems not to bother theoreticians of organized crime at all when Buscetta offered a radically different history than theory demanded. The criminological/sociological theory under scrutiny states that the high point in organized crime development both in the U.S. and Italy is La Cosa Nostra and/or the Commission of La Cosa Nostra, and it came about first and foremost to rationalize murder. This is where the history of organized crime leads. But for Sicily, the achievement came and went in a flicker of time. In Sicily we know without the slightest doubt that La Cosa Nostra restrained nothing. Despite the sociological road from feudality to bureaucracy, the Sicilians wouldn't behave as Anderson and others confidently predicted.

Listen to Buscetta's confession given in Palermo long before his appearance before the Permanent Subcommittee on Investigations. "Now, however, the informer Tomasso Buscetta, a longtime Mafia member, well acquainted with facts, broke the wall of Omerta. He did so for the purpose of destroying an organization that had <u>degenerated</u> into one of wild and unscrupulous criminals."(31) The key point stressed in Buscetta's story as outlined in the Tribunale di Palermo's 1984 indictment of 366 Mafia figures was that as an "uomo d'onore" it was all right to talk because the organization had lost honor. Buscetta, conceived of himself as standing for, what was artfully phrased, the "Permanence of Archaic Values." His confession ends with the assertion that he's just an old-fashioned guy operating in a world turned upside down by rapacity and venality; a Mafia world without values because it was without honor. Buscetta was most upset because the Mafia had perverted itself through violating certain organizational principles having to do with murder. As we know, Buscetta asserted Mafia principles demanded the regulation of murder: particular syndicates were not allowed to murder people in another syndicate's territory, not without permission in all cases from the "Commission, known as Cosa Nostra." The Mafia lost honor in proportion to its involvement in indiscriminate, unregulated murder.

But if the Sicilian Mafia was changing in the way Anderson and others speculated, than the strongest sign, the most important indication, would be discriminating regulated murder. How could one square the development of La Cosa Nostra and increasing Mafia warfare? The uncomplicated answer is there never was much of a connection between murder and criminal organization--organization did not reduce murder, and a diminution in violence at any particular time meant nothing concerning the structure of criminal organization. It appears that the qualities and history of La Cosa Nostra whether in the U.S. or in Sicily have been misread.

In Sicily the creation of La Cosa Nostra in 1963 was actually coterminous with the start of a war. It then became a cause for warfare which lasted for decades. The core of Buscetta's statements about the war, centers on control of "the Commission of the Cosa Nostra."(32) In 1963 the Cosa Nostra Commission "scattered" in the wake of battles between Palmeritan gangsters. Its ability to direct affairs a nullity. There were several other Cosa Nostra Commissions established in Sicily after 1963. Each was as ineffective in managing affairs and reducing violence as the first.

In fact, control of the Commission became a prize fought for by various Mafia groups. Cosa Nostra leadership, the Commission, conveyed status, hence it stimulated conflict.

The root of contemporary Mafia savagery lies in the competitive and treacherous relations among the major Mafia gangs both in Palermo and the traditional Mafia towns of Western Sicily. In 1984 Buscetta identified 38 Mafia gangs placing several in Palermo and locating others in towns such as Termini Imerese, Caccamo, Bagheria, Cinisi, Terrasini, Carini, Partinico, Borgetto, San Giuseppe Jato, Alcamo, Riesi, and Corleone which have had a Mafia presence for about a century.(33) The Corleone Mafia's leadership was held by Luciano Leggio, his assistants Salvatore Riina and Bernardo Provenzano, Leoluca and Salvatore Bagarella, and Michele Navarra.

Over the course of the past three decades the Corleone Mafia has exerted and expanded its power and influence against Mafia rivals primarily from Palermo. No doubt part of the Corleone Mafia's appetite for power and violence stems from its own recent history. Corleone is a small, bitterly poor, violent mountain town south of Palermo. The 1977 census reported a population of 11,057 making it one of the smaller Mafia "agro-towns."(34) From 1953 to 1958 Corleone lost about 1/66th of its population to murder. If sustained decade in and decade out, Corleone would have had as many actual murders in five decades as New York City experienced from 1866 through 1929.(35)

In some ways the Corleone Mafia was just another band of thugs in the long and dreary nightmare of that town's history. Some idea of life in Corleone can be gleaned by turning to the work of reformer, writer, and poet Danilo Dolci who interviewed peasants in Corleone and elicited a sort of reverie on murder. Killing was the context of their daily lives. Corleone natives measured the passage of time through the dates of different murders.

> The first contract they put out right after the war was on a guy named Cianciana. They got him on the square, maybe for political reasons. Michele Randisio and Zu Matteo Capra's son, the one with the bad hand, they disappeared. The bones were found when everybody was looking for Placido Rizzotto's body. It was in the same pit where they threw Donna Calorina Saporita's son. We could've loaded up a whole cart with all the bones from that hole. Angelo Gulotta's and Ciccio Navarra's brothers. . . . Here the criminals are protected by the government. Even when they catch your killer, ten days later he's out of jail. See, it's the government that's crazy. It's awful how many murders there've been since the

war. Here it never stops. . . . It's habit. It comes
natural, like killing a goat.(36)

The peasants provided an encapsulated history of Corleone's Mafia.
The first Mafioso remembered was Mariano Cuddetta "a real cutthroat,
stealing animals, killing people." He was followed by the "orphan Piddu
Uccedduzzu," then came Cicci Figatellu and subsequently Vincenzo
Crisciune. Each was surrounded by murder. Besides the people they killed,
Figatellu lost two brothers to other Mafia murderers, while Crisciune had a
son shot to death. Soon after Crisciune's son was murdered, he was too.
The last Mafia boss the peasants mentioned to Dolci was Doctor
Navarra--"He even made plans to have the president killed up there in
Rome," they said. And so it went, murder after murder until one of the
speakers turned to Dolci commenting "How the hell can I remember them
all?"(37)

Clearly enough, there is no reason to believe in a new "degenerate
Mafia," a category discovered by Buscetta to justify informing, as there is
none to accept the claimed correlation between violence and La Cosa Nostra
whether in the U.S. or Sicily.(38) The history of organized crime is far too
important to be written by those who see all its events through Commission-
tinted glasses.

* * * * * * * * * * * * * * * * * * *

NOTES

1. The literature on RICO is voluminous and I will make no attempt to cover
it all. One of the latest articles worth consulting is Patrick J. Ryan and
Robert J. Kelly, "Analysis of RICO and OCCA: Federal and State Legislative
Instruments Against Crime," Violence, Aggression, and Terrorism vol. 3
(1989); also see G. Robert Blakey and Brian Gettings, "Racketeer Influenced
and Corrupt Organizations (RICO): Basic Concepts--Criminal and Civil
Remedies," Temple Law Review Quarterly vol. 53 (1980); Whitney Lawrence
Schmidt, "The Racketeer Influenced and Corrupt Organizations Act: An
Analysis of the Confusion in its Application and a Proposal for Reform,"
Vanderbilt Law Review vol. 33 (1980); Andrew R. Bridges, "RICO Litigation,"
Georgia Law Review vol. 18 (1983); Robert P. Rhodes, Organized Crime:
Crime Control vs. Civil Liberties (Random House, 1984), particularly Chapter
8, "From Rico to Mail Fraud: Criminal Law Designed to Control Organized
Crime"; G. Richard Strafer, Ronald R. Massumi, Holly R. Skolnick, "Civil
RICO in the Public Interest: 'Everybody's Darling,'" American Criminal Law
Review vol. 19 (1982); and Denis Binder, "The Potential Application of RICO
in the Natural Resources/Environmental Law Context," Denver University
Law Review vol. 63 (1986). A most interesting review of RICO can be found

in the long statement "Additional Views of Commissioner Eugene H. Methvin," in President's Commission on Organized Crime, THE EDGE: Organized Crime, Business, and Labor Unions Appendix (Government Printing Office, October 1985). Of inestimable importance is Gerald Woods, Margaret E. Beare, "The RICO Statute: An Overview of U.S. Federal and State RICO Legislation," No. 1984-12, Research Division, Ministry of the Solicitor General of Canada.

2. Among the important work which set the tone and many of the categories found in the issues this essays addresses is Donald Cressey's "The Structure and Functions of Criminal Syndicates," in President's Commission on Law Enforcement and Administration of Justice, in Task Force Reports: Organized Crime (Government Printing Office, 1976), and his Theft of the Nation: The Structure of Organized Crime in America (Harper & Row, 1969).

There was a thoughtful scholarly attack launched during the 1970s in opposition to "classic" doctrine particularly at odds with the ethnic definition of organized crime. It was led by sociologist Joseph Albini in his work The American Mafia: Genesis of a Legend (Appleton-Century-Crofts, 1917). Other scholars such as Dwight C. Smith, Jr. The Mafia Mystique (Basic Books, 1975), and William J. Chambliss, On The Take: From Petty Crooks to Presidents (Indiana University Press, 1978) also made important contributions to the revisionist cause.

3. The 1980s has witnessed the most impressive and sustained "war on organized crime" in the history of American law enforcement. Thousands of racketeers including many of the most prominent leaders of long established crime syndicates have been successfully prosecuted. These accomplishments are of great significance and should never be underestimated, but that doesn't mean historical issues have been resolved.

4. United States District Court, Southern District of New York, UNITED STATES OF AMERICA, Plaintiff, against INTERNATIONAL BROTHERHOOD OF TEAMSTERS, CHAUFFEURS, WAREHOUSEMEN AND HELPERS OF AMERICA, AFL-CIO, et al., Defendants, "Declaration of RANDY M. MASTRO," 88 Civ., 9.

5. U.S. District Court, 9-10.

6. David Hackett Fischer, Historians' Fallacies: Toward a Logic of Historical Thought (Harper & Row, 1970), 43.

Fischer's work was aimed at professional historians who, he claimed correctly, committed logical and other philosophical errors time and again. It seemed that no historian, however prominent, escaped Fischer's attention. Of course historians are not the only ones to either write or invoke history to explain current phenomena, nor are they the only ones who must be called to account for mistakes.

7. Lonardo Interview by Special Agents Richard B. Hoke, R. Gerald Personen, CV 183A-1129 Sub H--10/28/83, FD-302.

8. U.S. Department of Justice, Criminal Investigative Division, Organized Crime Section, Organized Crime Intelligence and Analysis Unit, "Chronological History of La Cosa Nostra in the United States, January 1920--August 1987," (October 1987) included in U.S. Senate, Committee on Governmental Affairs, Permanent Subcommittee on Investigations, Hearings, ORGANIZED CRIME: 25 YEARS AFTER VALACHI, (Government Printing Office, 1988), 294. Further citations under Intelligence and Analysis Unit.

9. Fischer, 267, 270.

10. Intelligence and Analysis Unit, 295.

11. New York City Police Department, "Statement," in U.S. Senate, Committee on Governmental Affairs, Permanent Subcommittee on Investigations, Hearings, ORGANIZED CRIME: 25 YEARS AFTER VALACHI, 950.

12. Intelligence and Analysis Unit, 296.

13. Intelligence and Analysis Unite, 297.

14. Intelligence and Analysis Unit, 301.

15. Fischer, 62, 70.

16. Intelligence and Analysis Unit, 299-300.

17. Smith, Jr., 242, 324-27. Also see Smith, Jr., "Mafia: The Prototypical Alien Conspiracy," Annals 423 (January 1976).

18. See Humbert S. Nelli, The Business of Crime: Italians and Syndicate Crime in the United States (Oxford University Press, 1976), 179-184. In the decade or so since it was proven wrong, there have been no new sources identified which could conceivably have turned the issue around.

19. New York City Police Department, 900.

20. On the "fallacy of essences" see Fischer, 68.

21. Karl Popper's extended discussions on the philosophy of history are taken from Fischer, 68-69.

22. PSI, Hearings, 49.

23. PSI, Hearings, 49.

24. Luciano sailed to Italy on 9 February 1946 aboard the SS Laura Keene.

25. Speculation that the need for mobsters to "rationalize and routinize" murder has been at the foundation of much historical malarkey. On the one hand this was supposed to be the reason for the National Cosa Nostra, on the other it was described by mob informant Joe Valachi as apparently the major function of a New York Commission. In this instance Valachi was informed by another New York gangster that no member of La Cosa Nostra could be killed without the permission of the New York Council (or bosses). This localized version made sense to Valachi for several reasons perhaps foremost because he believed La Cosa Nostra was pretty much a New York affair. When asked by FBI interrogators, for instance, about the extent of "Causa [sic] Nostra," he answered there were about two thousand active members with "no more than a few hundred outside New York." See U.S. Department of Justice, Federal Bureau of Investigation, Report of James P. Flynn, "Joseph Valachi: Anti-Racketeering," Bureau File No.: 92-4282; Field Office File No.: 92-1459, 11 December 1962, 25.

26. See Dean C. Tipps, "Modernization theory and the Comparative Study of Societies: A Critical Perspective," Comparative Studies in Society and History, (1973). Tipps noted "many processes of change may be incompatible with others, and that because of differences in timing and initial setting processes of institutional change associated with modernization in one context need not be recapitulated in others. . .", 221. The relevance of this will be clear in the discussion of the Sicilian Cosa Nostra.

27. Robert T. Anderson, "From Mafia to Cosa Nostra," American Journal of Sociology, 307.

28. Anderson, 307.

29. Anderson, 309.

30. Giorgio Chinnici and Umberto Santino, L'omicidio a Palermo e provincia negli anni 1960-1966 e 1978-1984 (Centro siciliano di documentazaione "Giuseppe Impastato" e dall'Istituto di statistica socialie deel'Universita di Palermo, 1986).

31. Tribunale di Palermo, Ufficio Istruzione Dei Process Penali, Mandato Dicattura, Palermo, Sicilia, 29 Settembre 1984, 116 (my emphasis).

32. Tribunale di Palermo, 131. As reported by the Palermo Court, Cosa Nostra is a ruling clique composed of certain Mafia leaders trying to control the overall affairs of Mafia gangs in Western Sicily.

33. Tribunale di Palermo, 121-128.

34. The census figures used are in Hammond Inc., <u>Medallion World Atlas</u> (1977), 35-36.

35. See the data in Haven Emerson and Harriet E. Hughes (eds.), <u>Population, Births, Notifiable Diseases and Deaths, Assembled for New York City, 1866-1938</u> (The Delamar Institutue of Public Health, Columbia University, 1941).

36. Danilo Dolci, <u>Sicilian Lives</u> [translated by Justin Vitiello] (Pantheon Books, 1981), 177-79.

37. Dolci, 188.

38. The most important historical question to emerge from Buscetta's statement is not that of a Commission or of Luciano's influence, it is instead organizational change associated with vastly increased financial power. The Mafia controlled so much money its relationship with politicians was radically altered as well as its ability to "negotiate with the national and international financial world." See Ezio Mauro, "A New Mafia," in <u>Notizie Dall'Italia</u>, No. 8, November 1982, 5.

The Mafia faced the problem of "allocating this large 'surplus value' from heroin, sums of money vast enough to upset the economy of one of our Regions," journalist Mauro wrote. This brought about new and extremely dangerous connections between the Mafia and the Sicilian banking world. Mauro cites the rapid growth of Sicilian banking in comparison to Italian banking in general during the critical years of Mafia drug trafficking and quotes part of a recent report by the Sicilian Branch of Magistura Democratica (a Judges' and Magistrates' Association) which stated the subsequent proposition:

> . . . the growing weight of banks in the Sicilian economy, the common interests shared by Banks and Mafia businesses, the selection banks make in granting credits or in financing the strangest and unknown organizations, places the Banking System itself at the centre of financial manoeuvres, making them the load bearing pillar of the entrepreneurial activities of political-Mafia groups. Mauro, 6.

PART II:
ORGANIZING THE DRUG TRADE

Chapter 3
Organizing the Cocaine Trade in Progressive-Era New York

Published in <u>Criminology</u> (1979)

In the social scientific literature concerned with organized crime, historical claims, judgments, assumptions and opinions are rampant. Some of these have even been worked into a history of sorts, that is most notable for its reliance on undocumented popular sources and the unsubstantiated memoirs of celebrated informers and their ghost writers. Perhaps the most telling features of this history are its devotion to the ideas that an alien conspiracy has run organized crime for the last five decades, and that organized crime's development has been an inexorable march toward centralization and bureaucratization.(1) So compelling are these features that almost surreptitiously the very term organized crime has been carelessly transformed into meaning <u>the</u> monolithic organization of criminals. To write about organized crime is to saddle oneself with the outline of the ineluctable drive toward consolidation and confederation. This conundrum has recently moved several historical researchers to abandon the term organized crimes altogether in favor of the term illegal enterprises in their efforts to construct a responsible history. One example is Mark Haller who notes that "at the heart of what is often meant by organized crime are types of enterprises that sell illegal goods and services to customers: gambling, prostitution, bootlegging, narcotics and loansharking." Researchers, Haller advises, should first "identify a type of criminal activity" and then explore the questions of organization and coordination in a variety of historical settings.(2) Haller's modest proposal is a call for work that will build the historical superstructure without which contemporary work will remain attempting to prove a foregone conclusion.

To anticipate a rejoinder let me state that the fault for false historical claims, etc., is equally to be shared by the historical establishment which until recently has not considered the history of crime an important topic. This has meant that a diligent search for primary sources has hardly been mounted. In this way, the historians' neglect matches the "myopic research vision" commented on by William Chambliss in a short critique of the self-imposed research limits established by sociologists and criminologists interested in organized crime. For both disciplines "the data are there."(3)

Moving from organization to enterprise is something of a shock, however. For if the historical work dealing with the development of the supermob is shoddy, the history of the various activities identified with organized crime is almost nonexistent. Take narcotics for example. There is simply no acceptable historical work dealing with the history of narcotics trafficking in the United States. And even those works which deal with narcotics and organized crime in contemporary America, which attempt to provide some historical background, are terribly confused.

This essay will explore the drug trade as organized within a segment of New York City's underworld during the 1910s. Among the major reasons for selecting this period is that it was clearly the time during which national concern focused on narcotics resulting in such legislation as the Harrison Act. In trying to come to grips with a "new social problem" attention was centered on both a particular drug, heroin, and the population of users, not distributors. In response, and as would be expected, a supposedly scientific literature was established detailing the tragic and disgusting destruction of thousands of lives through addiction. An important part of this emerging concentration was the view that drug use meant complete and total abandonment of will and purpose. Drug use signalled, moreover, a previously weak and disorganized character formation in the user. Addiction in this sense only proved what had been obviously believed before--that vast numbers of the urban poor were character deficient. Unable to successfully complete, willfully unpredictable, at best casually disciplined, drug addiction was the capstone of failed lives.(4) Within this litany of degradation, the only consistent picture of the drug distributor to emerge was fiendishly drawn. The distributor was the dope pusher, poisoner of countless young lives, destroyer of the weak. Clearly, racial, sexual, class and ethnic stereotypes abounded in this literature and have been carried forward in innumerable studies. With remarkable ease, Progressive-Era formulations have provided the intellectual bases for many of the contemporary conspiracy theorists. Needless to add, the "dynamics" of the drug trade during the Progressive Period have to be seriously reexamined outside the confines of Progressive morality. Earlier it was suggested that "the data are there" for both the historian and sociologist of organized crime. In this particular case, the data describing the illicit narcotics enterprises are the reports of an American-Jewish self-defense organization called the New York Kehillah.(5)

The Communal Response

With its roots in Jewry's European past, the New York Kehillah set out under the leadership of Judah L. Magnes to order Jewish education, philanthropy, labor and morals. While the trend toward a Kehillah had been building for some time, the immediate catalyst in the development was a charge of Jewish criminality leveled by New York's Police Commissioner,

Theodore A. Bingham, in the late summer of 1908. Bingham's statement which appeared in the <u>North American Review</u> held that around fifty percent of the criminal classes in New York City were Jews.(6) This charge of overwhelming Jewish criminality was reinforced a year later with the appearance of three famous articles in <u>McClure's Magazine</u> calling attention to Jewish involvement in organized criminal activities. The conspicuous concern with crime during this period marked a new and long season of reform for New York City. Fed by muckraking articles, government reports, published investigations, and the findings of a special Grand Jury, the public became privy to the details of a vast complex of prostitution and gambling. And then, in 1912, when the recurrent absorption with lawlessness and immorality appeared to have run its course, the murder of the notorious gambler, Herman Rosenthal, revitalized public concern with Jewish criminality especially its manifestations in organized illicit businesses.(7)

In the arena of public morals, this was the situation confronting the Kehillah. Responding to this crisis, the Kehillah established a Bureau of Social Morals in the late summer of 1912. The Bureau was to be an anti-crime unit staffed by a number of private investigators whose province was primarily the six police precincts of New York's Lower East Side. The Bureau's task was to drive crime out of the Lower East Side by gathering as much hard evidence as it could through private investigations and then working cooperatively with Mayor William J. Gaynor and his major police officials. Magnes and the Bureau's counsel, Harry W. Newburger, brought to the Mayor lists of criminal establishments they wanted closed: the Mayor was to make sure that the police raided the offending places. To back up the police, the Bureau had an attorney attend those criminal court cases affecting the Lower East Side and Jewish organized criminals.(8)

The effectiveness of the Bureau of Social Morals was sharply reduced after its first year of operation when Mayor Gaynor, whose support was absolutely central, died in September 1913. Without Gaynor the Bureau's political influence was severely curtailed. Quiet for a number of years, the Bureau came back to life for only a short period of time in 1917, when Magnes tried to revitalize it by garnering money from a number of wealthy Jewish philanthropists. Receiving some financial aid, the Bureau concentrated one last time on uncovering crime among New York's Jewish population.(9)

Up until 1917, the Bureau's reports had been rather discursive documenting a wide range of criminal activities and influences. But in the late spring of 1917, during the last phase of Bureau activity, it centered on what was considered to be the most pressing criminal problem in the Jewish neighborhoods--narcotics. There is no mistaking the seriousness of the Bureau's perception of the problem as shown by its statement prefacing the reports. Jewish drug dealers, it is remarked, "were creeping into the homes of the East Side, Bronx, Harlem and Brooklyn with their poisonous drugs and if this thing is not curbed . . . tens of thousands of our men and women will be hopelessly addicted to the use of cocaine and opium." It was added that "the mind and body of Jewry" were being "attacked and poisoned" by fellow Jews which for the Bureau was the cruelest irony of all.(10)

Interestingly, the knowledge of Jewish involvement in the illegal drug trade did not bring on anything resembling the earlier nativist articles and accusations. The issue of Jewish criminality which had been fed by the revelations of Jewish gamblers, prostitutes, pimps and madams had obviously expired by 1917. In fact, Jewish crime had become so dead an issue by the second half of the decade, that there is hardly a reference to be found in either the contemporary or historical literature (except of course, for the Bureau's primary material) on crime or American Jews which discusses Jewish criminals and the drug trade during the 1910s.(11) The discussion which follows is based upon the information contained in the Bureau's reports on narcotics. There was no final comprehensive report on the menace of drugs, and there was no particular organizational scheme evident in the material. The investigators' purpose was to document Jewish involvement in narcotics and then to turn their evidence over to the appropriate municipal agencies, which this time did nothing. The reports are a day-by-day account of the information uncovered by the investigators with little attempt to coordinate one day's findings with the next. But within the mass of material there are data concerning age, ethnicity, criminal occupations, kinship, arrest records, geographical locations of criminal hangouts, function in the narcotics trade, and membership in particular mobs when it applied for 263 identified drug dealers. There are some things to be noted about the data, however. The information is not consistent for all members of the sample; for some, all that was reported was name and age, for others, name and arrest records, and for still others, nickname and location of hangout. The material has been arranged to answer three interrelated questions: first, the personal background of the dealers; second, the structure of the illicit narcotics trade; and third, the structure and stability of the mobs involved in the trade. Mobs, hereafter called combinations, are defined as collectives having at least three apparently equal partners. One final point concerns the drugs themselves: with very few exceptions all the dealers were in the cocaine business; probably one of the "unintended consequences" of the Harrison Act and the fixation on heroin as well as the severe disruptions of settled patterns of international trade caused by World War I.

Personal Background

Given the Bureau's purpose and concern, ethnicity is the first aspect of personal background to be examined. Ethnicity has been determined by either the Bureau investigators' explicit statements or by my judgment based on the reported names. For instance, "Italian Frank" and "Augie the Wop" were counted as Italians, and Bernstein, "Yeshiva Bocher," and Issie Katz as

Jews even though the investigators did not so identify them. Because of the uncertainty in such a method, ethnic identity has been conservatively estimated with 263 criminals ethnically identified. There are 83 Jews, 23 Italians, 8 Irish, 5 Blacks and 3 Greeks.

Even at this early stage two important points seem fairly clear: Jews were disproportionately over-represented in the cocaine trade; and it was not an exclusively Jewish business. This latter point is more apparent when consideration is given to the Bureau's desire to ferret out Jewish criminals probably to the exclusion of some others. The three smallest groups are only of passing interesting because of their very limited involvement in that segment of the underworld investigated by the Bureau. The Greeks were members of a small retail operation located in a Greek restaurant on Sixth Avenue in Manhattan. Unlike the Greeks, the five Blacks were independent dealers. Griff Richardson sold narcotics out of a pool parlor on West 135th Street in Harlem. The four remaining Black dealers were Dick Green, "Blackie," Bob Kemp considered a large dealer whose operation was located in a restaurant on Lexington Avenue in Harlem, and Dick Brown who worked the Lower West Side of Manhattan. Concerning the Irish dealers, the Bureau's investigators provided no information beyond name. The Italians, as will be discussed later, played a variety of roles in this predominantly Jewish underworld.

Two other categories of personal background are sex and drug use. There were four women dealers among the 263 criminals. None of them worked alone.(12) Three of them were members of two combinations held together by marriage and kinship. Concerning drug use the Bureau stated that 57 of the dealers were users. However, users were relatively more prevalent in simple partnerships (two individuals) than in combinations. Out of the 38 people constituting the nineteen partnerships found in the sample, 26 percent were users, while only 9.5 percent of combination members were similarly inclined. Not much should be made of this, though, as it was a rare partnership in which both members were reported as users. The bulk (two-thirds) of the users came from the ranks of independent and small dealers.

Another important area in the personal backgrounds of the dealers concerns their involvement in other illicit activities. Apart from narcotics, twenty different criminal enterprises were conducted simultaneously with drug dealing. The twenty collateral activities fall into five fairly distinct types of enterprises: stealing; gambling; vice; criminal management which includes fencing of stolen goods, political fixing, and managing or owning of criminal establishments; and extortion. The number of drug dealers found participating in the five major collateral criminal occupations is: stealing 50; gambling 7; vice 32; management 32; extortion 21.

The collateral occupations are exceptionally significant as they indicate the complex interplay of criminal activities revealing, for example, the relationships between drug dealers who were pimps and the groups of retailers who primarily sold to prostitutes. The range of collateral activities also establishes that drug dealers were more likely to be criminal generalists not specialists and that drug dealing was typically part of an expanding criminal repertoire. These last points are, of course, important factors in the social structure and stability of combinations.

Some confirmation of the spread of criminal activities is found in the Bureau reports which mentioned arrest records. Slightly more than 100 of the dealers were found to have been arrested for offenses ranging over twelve different crimes including murder, stealing, burglary, armed robbery, pimping, gambling, strikebreaking, prostitution and narcotics violations. By far the most common arrests were for drugs with various forms of stealing a distinct second. About one-fifth of the dealers arrested had been arrested at least once before.

Drug dealing during this period was clearly the kind of enterprise that demanded some sort of criminal maturity. It was not the sort of activity readily engaged in by juvenile delinquents or exceptionally youthful criminals, according to the reports. Age was one of the areas in the personal background of the criminals more extensively recorded. The ages of over 86 percent of the dealers were given and they ranged from 45 to 18 with the average 28. Although the largest clustering falls in the years 25 through 29 with 99 of the dealers in this group, it is striking that 84 others were over 30. Only 44 of the dealers were younger than 25 and almost half of this group were 24 years old.

The drug--more properly cocaine--industry uncovered by the Bureau of Social Morals in the summer of 1917 was staffed by a population that was heavily Jewish, although not exclusively so, and overwhelmingly male. At least one-fifth of them were habitual drug users and more than one-third of them had been arrested at least once. It follows that they were knowledgeable about the operations of the city's criminal justice system, strangers neither to the criminal courts, the city's jails, nor the legion of bondsmen (and women) and lawyers who worked the system. In the main, this population was criminally mature involved in a multiplicity of illegal activities which aided them in developing the contacts and capital resources to successfully deal in drugs.

The Structure of Snow

As this population was essentially concerned with the distribution of cocaine it roughly divided itself into importers, wholesalers and retailers--the structure of the industry as a whole. The division, however, was neither neat nor equal with the vast majority of dealers engaged solely in retailing. Disposed to seizing all sorts of criminal opportunities, the 234 dealers for whom there is information about their marketing habits, actually comprised seven different and overlapping categories. There were 2 dealers who were exclusively importers, 10 wholesalers, and 173 retailers. But there were also 4 individuals who were both importers and wholesalers, 2 importers and retailers, 27 wholesalers and retailers and 16 who were engaged in all phases of marketing.

The only two individuals who were exclusively importers were "Uncle" Joe DeGorizia, a well-known figure in the underworld of Paterson, New Jersey, and Dave Lewis equally famous in the underworlds of both Manhattan and Brooklyn. As a member of the "Kid" Springer Gang, Lewis usually could be seen in a restaurant on Pitkin Avenue in the Brooklyn neighborhood of Brownsville. There is no other information about their activities or where there drugs were imported. There is, however, enough material on the 22 other part-time importers to roughly indicate points of cocaine entry. For instance, "John the Sailor" who participated in all three distribution categories worked ships that berthed in New Orleans where he presumably received cocaine. Other areas noted were Buffalo, New York, Philadelphia, various New Jersey ports, and Canada.

The material on wholesaling is more substantial. Most wholesaling was carried out by non-specialists. Additionally, over two-thirds of this group were members of combinations. Outside of a few major independent wholesalers, it seems that this aspect of the trade was dominated by the combinations, although only one combination confined itself totally to the wholesale trade. Among the ten individuals who specialized in wholesaling, four formed a single combination which was headquartered in a saloon on West 45th Street in Manhattan. But the four marketing specialists were still criminal generalists: three were pimps while the fourth was a bookmaker.

Probably the most important independent wholesaler was Al Lampre, known as "Cockeye Al of 14th Street." Lampre supplied at least ten dealers whose reach was city-wide. Additionally, he franchised cocaine selling it to Eddie Friedman, Hyman "Nigger," and Jimmie McDonald who retailed it and then returned part of their profit to Lampre. Friedman sold in downtown Brooklyn, Hyman "Nigger" on Avenue B in the Lower East Side, and McDonald along the Bowery including Harry Callahan's political club. Among the retailers who bought cocaine outright from Lampre were Sammy Klein, Joe "Stagger," Sammy Cohen, and partners "Farge" and "Dago Jimmy," and Charlie Palmer and Mike Goodman. One of Lampre's sources of supply was identified. He was Harry Witt who was located in a saloon in West New York, New Jersey. After Lampre, the most significant independent cocaine wholesaler was Ike "Pheno" Brown who could usually be found in Wolpin's restaurant on Broadway in the theatre and night club district. Brown's business was structured in the same manner as Lampre's only on a somewhat smaller scale. Those dealers noted as Brown's customers dealt only in Bowery hotels and pool parlors. Although the investigators did not directly mention his supplier(s), they did suggest that Abe Kutner of Buffalo was probably one. Another wholesaler who adds to the geographical picture of the cocaine trade was Johnny Daly who lived and worked in Schenectady, New York. Buying from Daly was a retailer, Tony Abato, who brought the narcotics back from up-state New York and sold it along the Broadway theatre district.

Two other areas outside of New York City were also mentioned as places where New York wholesalers worked: Boston and Philadelphia with the latter being particularly important. The criminal connections between New York and Philadelphia, which have never been explored, were lucrative and well-established. Besides the already noted importation of cocaine in Philadelphia, a number of the New York combinations also sold there. One example is the Britt Brothers' combination which worked both the Lower East Side and Philadelphia. Another New York group prominent in the Philadelphia underworld was formed by Abie Cohen, "Desperate Little Yidel," "Fatkie" and "Little Jimmy."(13) New York's largest drug combination, named the "Silver and Ream" one, was also represented in Philadelphia. Its man there was Johnnie Burt who had an extensive criminal background in New York, Chicago, and sold cocaine along Noble, Vine and Race Streets in Philadelphia. Silver and Ream also had a partner working in Boston. There were at least six wholesalers representing several combinations working in Philadelphia as well as two in Boston.

The predominant activity in the drug trade was, of course, retailing: seventy-four percent of the 234 dealers whose marketing habits are known were strictly in the retail trade. If the part-time retailers are added, retailing encompasses 93 percent of the sample. The characteristics of retailers were obviously those of the dealers in general. There are, nevertheless, some points to be made. There were at least nine different types of places in and on which retailers conducted business: parks, drug stores, hotels, restaurants, saloons, pool parlors, theatres, public buildings and certain street corners throughout New York City. To determine geographical concentrations of cocaine retailers a list was drawn of the 129 specific addresses found in the Bureau's reports. The two areas of greatest concentration were (roughly) the Lower East Side with 44 percent of the addresses and the Broadway district with 15.5 percent of them. Other areas which showed clusters were Harlem, several sections in Brooklyn and the Mott and Mulberry streets area in Manhattan which was adjacent to the Lower East Side.

In addition to location, the investigators reported on four different types of retail consumers: prostitutes, actors and actresses, soldiers and sailors, and newspaper people. Those who dealt either solely or primarily with prostitutes were partners Mike Butler and Bill Ronan, "Smiling Frank" and Pete Braga, Charlie Palmer and Mike Goodman, Hymie Sindler and "Dago Alec," and Jim Malloy and Tony Cousine. Besides the partnerships, two combinations' retail trade was oriented toward prostitutes. One was the Klein combination, the other was formed by Max Frank, Sam Freidman and

Sam Goldstein. The material on Goldstein reveals something of the web of business relations between dealers and vice entrepreneurs. Goldstein had a business and personal relationship with a bondswoman, Winnie Meyer, who had earlier been a madam. It seems that she both solicited customers for Goldstein who sold only to prostitutes, as well as protected him from prosecution using her political influence with local magistrates.

Those whose prime market was actors and actresses were the LeRoy Combination and two independent retailers, Johnny Fox and Franklin. Both Fox and Franklin worked in the same establishment, the New York Victoria Hotel on West 47th Street right off Broadway; there is no evidence that they worked together. While the connections between show people and vice including narcotics was if not widely known, at least deeply believed,(14) a similar exchange relationship involving the military was not. Nevertheless, soldiers and sailors furnished an extensive market for some of the dealers. All of the reported selling to the military took place in Brooklyn at the Brooklyn Navy Yard and the 47th Regiment Armory. One of the dealers specializing in this area was "Skelly" whose operation employed at least a dozen men as sellers. The other dealers were Max Gardner and "Johnny Spanish" who were among the most notorious criminals in New York.(15) Garnder and "Spanish" were partners in this aspect of the cocaine trade, as well as fellow members in the "Kid" Springer gang. Another member of the gang was importer Dave Lewis, although there is no evidence that Lewis worked with either "Spanish" or Gardner in cocaine. To add to the organizational confusion, Gardner was also reportedly in a separate drug combination with his three brothers.

The last consumer group was newspaper people who played several roles in the cocaine trade. In addition to being consumers, there were two unnamed "newsboys" who were retailers selling exclusively to streetwalkers. One peddled both newspapers and narcotics in front of "Pheno" Brown's headquarters, Wolpin's restaurant. Newsboys also functioned as spotters and contacts for a number of dealers. Several newsboys worked as sellers for Joe "Rocks"; and lastly, one of the dealers in the sample is "Dutchy," a twenty-six year old "newsboy" who worked for Charlie Young a wholesaler.

The particular marketing specialties for some of the dealers may be explained by geographical proximity. For example, two of the dealers who sold to prostitutes lived in the same building at 31 East 14th Street. On that same block were several different female institutions including the Little Mothers Aid Association, the Little Sisters of the Poor, and the Hebrew Technical School for Girls. Joe "Rocks" also lived on that block and it is notable that the Newsboys Club was only a short distance away on the corner of East 11th Street and Second Avenue.

Cocaine was imported, wholesaled, franchised and retailed: it moved from South America to New Orleans, Canada, Buffalo, Philadelphia, West

New York, New Jersey, and so on. It traveled back and forth from Broadway and the Forties to the Lower East Side, Harlem, Brooklyn, back to Philadelphia and also Boston. It slid up and down the Bowery and Second Avenue and Third Avenue, across 14th Street where it circled Tammany Hall, then slipping down toward Mott and Mulberry Streets ending up perhaps at the Essex Market Court on First Street and Second Avenue (where it was sold, not used for evidence), or at the Odd Fellows Hall at 98 Forsythe Street.(16) It was sold many times over by a variety of dealers (who were also at times, consumers) to show people, the military, newsboys, prostitutes, Jews, Italians, Blacks, and so on. It was traded in movies, theatres, restaurants, cafes, cabarets, pool parlors, saloons, parks and on innumerable street corners. It was an important part of the coin of an underworld that was deeply embedded in the urban culture of New York; supporters as well as exploiters of the myriad establishments that made-up the night life of the city. And finally, cocaine was something of a bonding agent bringing criminals together in a variety of ways.

Combinations

There were 22 combinations central to the cocaine trade; the largest probably had seventeen members,(17) the smallest only three, the minimum number of partners defining a combination. Ten of the combinations had only this minimum size; five had four and two others had five. The last four combinations had six, seven, eight and nine members. The average size of a combination was about five members, although if the largest and most atypical one is discounted, the average dips to just over four members. The combinations were heavily Jewish: 85 percent of the ethnically identifiable members were Jews. The only other identified group represented were Italians who made up the remaining 15 percent. The ten Italians were distributed in seven different combinations, always in partnership with Jews. There were three combinations in which Italians and Jews reached parity, and thirteen others which were totally Jewish as far as can be determined. Both the predominant Jewish cast of the combinations, and the evidence of some inter-ethnic cooperation are notable.(18)

The marketing preferences of 85 combination members is known and they ranged in the following manner:

TABLE 1
MARKETING PREFERENCES OF
COMBINATION MEMBERS IN NUMBER
AND PERCENT ALONG WITH COMPARABLE
PERCENT OF THE REMAINDER OF THE SAMPLE

	Wholesale	Retail	Import Wholesale	Wholesale Retail	Import Wholesale Retail
Number from Combinations	4	42	2	22	15
Percent from Combinations	4.7	49.4	2.4	25.9	17.6
Percent from Remainder of Sample	4.0	87.9	1.3	3.4	0.67

Combinations were obviously much more deeply involved in the wholesale and import aspects of the trade than the rest of the dealers. In fact, adding the part and full-time wholesalers together about half the combination members did some wholesaling compared to 9.4 percent of the other dealers. In importing, the corresponding figures are 20 percent for combination members to about 5 percent for the rest. Retailing was still, however, the predominant activity even for combinations with eight of them exclusively in the retail trade. While importing and wholesaling were more likely to have been carried out by combinations in the ratio 3:1, it is also clear that combination members were somewhat less likely to have been marketing specialists than the rest of the dealers. This means that the social structure of these organizations was not typically marked by any particular, enduring division of labor structured around set marketing tasks.

Not only were combinations usually small groups engaged in changing marketing activities suggestive of informal organizations, but ten of them were at least partly forged out of kinship ties. Already mentioned are three, LeRoys, Kleins, and Gardners, which were nothing more than kin groups. In five of the remaining seven kin-combinations, at least half of the members were related. For example, one was composed of the three Newman brothers along with Charlie Straus and "Joe the Wop"; another featured two sets of brothers working with two unrelated dealers. One other example is the combination formed by Irving "Waxey Gordon" Wexler, his brother, known as Harry Irving, and two others.

Small loosely structured, often kin-centered, combinations were also notable for their lack of both stability and solidarity. Members had a pronounced tendency to divide both their time and ties among other criminal groups. For instance, in the Newman combinations one of the brothers divided his loyalty by working for another mob managing a crap game in Harlem. Obviously, collateral activities reduced the commitment to any particular combination. This same sort of division or splitting also took place within the cocaine trade as select individuals moved from one drug combination to another without forfeiting membership. This kind of situation was alluded to earlier in the discussion of the drug partnership of Max Gardner and "Johnny Spanish" which co-existed with the Gardner brothers' combination as well as Max's involvement with the "Kid" Springer gang. Perhaps the best example of this movement is in the activities of Irving Wexler who participated in five different cocaine combinations at the same time. One was formed with his brother and two others to work Harlem; another operated in Philadelphia; a third was composed of "Little Simon," Hershel Chalamudnick and Wexler and worked out of a tea room on Grand Street in the Lower East Side; the fourth was founded by Wexler, Hymie Fishel, and "Mahlo" and was headquartered in the Odd Fellows Hall at 98 Forsythe Street also in the Lower East Side; the last was made up of Wexler, his oldest associate "Jonsey the Wop," and six others working around 14th Street. Wexler's many pursuits were independent ventures in the sense that no other members of the five combinations including his brother crossed the same combination lines. Various of them did, however, cross other lines enjoying parallel memberships both in drug combinations and collateral mobs. Several examples follow: "Jew Murphy," one of Wexler's partners in Philadelphia was also a partner in another combination in New York; Hymie Fishel divided his time between a Wexler group and another combination headquartered in a saloon on Madison Street; Jack Pipes, the third man in Wexler's Philadelphia combination, was also in a separate group that worked Philadelphia; and Hershel Chalamudnick, an associate in the tea room combination was the fence for several burglary mobs operating around Coney Island in the Borough of Brooklyn. Concentrating on just cross memberships in drug combinations, leaving out the examples of membership in collateral enterprises and organizations, there were twelve combinations affected. This means that over half of the drug combinations contained members who had working interests in other drug combinations.

These groups had few of the attributes of "formal organizations."(19) This does not mean, however, that they functioned poorly. On the contrary, they were among the vehicles through which drugs were processed and fortunes made. In fact, their informal structure and probably short life-span were exceptionally responsive to the necessities of the drug trade. First of all, entry into the trade was fairly simple involving few costs beyond the initial capital investment, few contacts in the area of supply, and hardly any organization for distribution. But there was one catch: everything was contingent on overseas suppliers who could not be controlled by the American entrepreneurs. It would have been foolish to stake one's criminal career around a particular combination given the chances that there would be nothing to sell. Unable to control the delivery of the commodity, it made little sense to surround oneself with a formal marketing organization.

The field was lucrative, but demanded entrepreneurs who were flexible, with numerous contacts, able to raise capital at unexpected times and to pull together a small organization with little effort. Nothing adapted itself quite so well to these circumstances as ties of kinship which could be invoked to form the nucleus of a more-or-less spontaneous combination. Because they did not compete for limited sources or for a limited market, combinations were complementary. Dependent on the inefficiencies of supply, combination structure was exceptionally loose.

Discussion

At this point one may legitimately ask what the preceding rather detailed analysis of a highly selective segment of New York's underworld has revealed. First, of course, it is a corrective of the common notion that organized crime has been some sort of monolith marching in lock step toward ever increasing centralization. Quite the contrary, at least in drugs where decentralization was clearly the norm. The drug industry uncovered by the Bureau of Social Morals was fragmented, kaleidoscopic and sprawling. It was organized and coordinated not by any particular organization, but by criminal entrepreneurs who formed, reformed, split and came together again as opportunity arose and when they were able. In 1917, after several decades of drug trafficking there was little necessity for either centralization or indeed bureaucratization. Decentralization has probably always been the norm in the distribution end of the narcotics trade. Consider the similarity between the trade in 1917, and the contemporary scene described recently by Patrick Murphy the former Police Commissioner of New York. Murphy states:

> The illegal drug industry today can more accurately be compared structurally to the garment industry. Many sources of raw materials exist. Many organizations, large and small, buy and process the raw materials, import the product into this country, where it is sold to and processed and distributed at retail by a host of outlets, some large and small, chain and owner-operated . . . Organization in the drug business is largely spontaneous, with anybody free to enter it at any level if he has the money, the supplier and the ability to escape arrest or robbery.(20)

The analysis also places the narcotics trade during the second decade of the twentieth century within the broader context of a multiplicity of illegal enterprises engaged in by organized criminals. It suggests that such criminals were in reality criminal justice entrepreneurs acutely responsive to a broad panoply of activities which often bridged the gap between illegal enterprises and positions within New York's criminal justice bureaucracies. It also details the manner in which criminal careers were structured--not within a particular organization but through an increasing web of small but efficient organizations.

Moreover, the discussion broadens our understanding of ethnicity and organized crime. Whether or not the Jewish underworld revealed by the Bureau was a large or small part of New York's underworld, it did exist and was significant. Also notable is the evidence of inter-ethnic cooperation which clearly suggests that at times parochialism was overcome by New York's criminals. Additionally, the study validates Haller's call for a logic of historical investigation into organized crime. Especially for those works dealing with the social structure of organized crime, it is imperative to begin with an investigation of a particular criminal activity or enterprise and then move to considerations of organization and coordination. Only by approaching the problem from this perspective can one appreciate the interplay between the structure of the narcotics industry as a whole, and the structures of the partnerships and combinations marketing narcotics.

Finally, I strongly suggest the importance of viewing organized crime within its urban historical context. Replete as this study is with examples of the mesh between drug dealing and such urban structures as saloons, pool parlors, restaurants, and theatres, as well as those aspects of New York's economy which recruited and supported prostitutes, actors and actresses, and newsboys, it is clear that illegal enterprises are highly responsive to all manner of urban configurations. The final argument then, is that to understand organized crime one must understand urban political economy: to understand the history of organized crime one must venture into the study of urban history.

* * * * * * * * * * * * * * * * * *

NOTES

1. One of the best examples is Donald R. Cressey, Theft of the Nation: The Structure of Crime in America (Harper & Row, 1967).

2. See Mark H. Haller's, "The Rise of Criminal Syndicates," (unpublished manuscript, 1975), and "Organized Crime in Urban Society," Journal of Social History (winter 1971-1972).

3. William J. Chambliss, "On the Paucity of Original Research on Organized Crime: A Footnote to Galliher and Cain," American Sociologist (Febraury 1975), 36.

4. See Norman S. Clark, Deliver Us From Evil: Interpretations of American Prohibition (W. W. Norton, 1976), 217-226.

5. Arthur A. Goren, New York Jews and the Quest for Community: The Kehillah Experiment, 1908-1922 (Columbia University Press, 1970).

6. Goren, 25.

7. Theodore A. Bingham, "The Organized Criminals of New York," McClure's Magazine (November 1909); S. S. McClure, "The Tammanyizing of a Civilization," McClure's Magazine (November 1909); and George Kibbe Turner, "The Daughters of the Poor: A Plain Story of the Development of New York City as a Leading Center of the White Slave Trade of the World Under Tammany Hall," McClure's Magazine (November 1909).

8. The high point in the Kehillah's anti-crime drive came in 1913. In February of that year, Mayor Gaynor appointed Newburger the Third Deputy Commissioner of Police and agreed that in addition to Newburger's regular duties, "'everything affecting the First Inspection District would go through him.'" Accompanying Newburger as his secretary was Abraham Schoenfeld, the Kehillah's chief investigator. Newburger "ordered and led raids" while Schoenfeld was in charge of "one of the two roving 'strong arm' squads." Goren, 170-175.

9. Goren, 183-185.

10. Information on the Bureau of Social Morals was obtained from the Judah L. Magnes Archives, The Central Archives for the History of the Jewish People, Jerusalem, Israel. I first became acquainted with the material in the Magnes Archives through Goren's study noted above. In his note on sources he states that the Magnes Archives in Jerusalem "contain an outstanding collection of sources for the study of Jewish life in New York and Jewish communal politics in America from 1908 to 1922 . . . . The largest part of this material consists of the Kehillah's records which contain a wealth of sources on Jewish education, religious life, philanthropic organizations, industrial conditions and crime." Using Goren's citation for the special material on crime-MA (SP/125-SP/139--I wrote to the Central Archives for the History of the Jewish People, Jerusalem, Israel, and requested a microfilm copy of the almost 2,000 "case histories of Jewish criminals prepared by the Kehillah's chief investigator and based on information supplied by his informers and agents." Ms. Hadassah Assouline of the Central Archives was kind enough to fulfill my request.

11. Moses Rischin, The Promised City: New York's Jews, 1870-1914 (Harper & Row, 1962); Irving Howe, The World of Our Fathers: The Journey of the East European Jews to America and the Life They Found and Made (Harcourt Brace Jovanovich, 1976); Oscar Handlin, Race and Nationality in American Life (Doubleday, 1957); A. F. Landesman, Brownsville: The Birth, Development and Passing of a Jewish Community in New York (Bloch Publishing, 1971); I. Metzker (ed.) A Bintel Brief: Sixty Years of Letters

from the Lower East Side to the "Jewish Daily Forward" (Doubleday 1971); Harry Roskolenko, The Time That Was Then: The Lower East Side 1900-1914 (Dial, 1971); Joel Slonim, "The Jewish Gangster," Reflex (July 2, 1928); S. D. Hubbard, "The New York City Narcotic Clinic and Differing Points of View on Narcotic Addicition," Monthly Bulletin of the Department of Health, City of New York (February 1920); C. B. Pearson, "A Study of Degeneracy as Seen Among Morphine Addicts," International Record of Medecine (1919); Mayor LaGuardia's Committee on Marihuana, "The Marihuana Problem in the City of New York," in David Solomon (ed.) The Marihuana Papers (New American Library, 1966); and Herbert Asbury, The Gangs of New York: An Informal History of the Underworld (Alfred A. Knopf 1927).

12. For a discussion of female criminality in the Lower East Side during the 1910s see my "Aw! You Mother's in the Mafia: Women Criminals in Progressive New York," Contemporary Crises: Crime, Law, Social Policy (January 1977).

13. Most of the Philadelphia connections reported on by the Bureau seem to have originated with the activities of Benjamin "Dopey Bennie" Fein, one of the most important Jewish mobsters whose career was over a few years before 1917. A large number of the dealers who worked both in New York and Philadelphia had at one time been affiliated with Fein; "Little Jimmy" was his cousin. In a report filed by the Bureau's investigators in the summer of 1913, dealing with labor racketeering, it was noted that Fein also operated in Philadelphia as a labor thug. SP/134, Story #700.

14. A.F. Harlow, Old Bowery Days: The Chronicles of a Famous Street (D. Appleton, 1931), 403-407.

15. Asbury, 261-264.

16. The Odd Fellows Hall was the most popular meeting place and distribution center in the cocaine trade: seven combinations were found to congregate there in 1917. The building continued to serve as a narcotic center through the 1920s as a story in The New York Times, June 12, 1926, p. 17, indicates. On the preceding day, federal narcotics agents had arrested six men at the Odd Fellows Hall and confiscated $25,000 worth of cocaine, morphine and heroin. Unknown at the time, one of the men arrested was Irving Bitz who was, along with his partner, Salvatore Spitale, one of the area's leading narcotics dealers. For a disguised version of Bitz's career, see I. W. Halpern, J. N. Stanislaus and B. Botein, The Slum and Crime: A Statistical Study of the Distribution of Adult and Juvenile Delinquents in the Boroughs of Manhattan and Brooklyn (New York Housing Authority, 1934), 131-153.

17. There is some confusion in the reports on the size of the largest

combination known as the Silver and Ream Syndicate. In some reports it appears to have seventeen full partners, in others no more than five.

18. In the more-or-less traditional histories of organized crime, interethnic cooperation and multi-ethnic syndicates are supposedly a feature of the restructuring of organized crime that took place in the early 1930s. Prior to this time, the underworld was supposedly notable for its ethnic divisions and indeed wars.

19. See Peter M. Blau and W. R. Scott, Formal Organizations: A Comparative Approach (Chandler, 1962).

20. Murphy's statement is in Staff and Editors of Newsday, The Heroin Trail (New American Library, 1973), 187.

Chapter 4
European Drug Traffic and Traffickers
Between the Wars

Published in <u>Journal of Social History</u> (1989).

"You don't seem to approve of drug-taking."

"Approve!" Mr. Peters stared, aghast. It is terrible, <u>terrible</u>! Lives are ruined. They lose the power to work yet they must find money to pay for their special stuff. Under such circumstances people become desperate and may even do something criminal to get it. I see what is in your mind, Mr. Latimer. You feel that it is strange that I should have been connected with, that I should have made money out of, a thing of which I disapprove so sternly. But consider. If I had not made the money, someone else would have done so. Not one of those unfortunate creatures would have been any better off and I should have lost money."

Eric Ambler, <u>A Coffin For Dimitrious</u> (1939)

There is no satisfactory comprehensive history of the illicit traffic in narcotics (opium, morphine, heroin, and cocaine) in the twentieth century. Some portions of this complex history have been well recounted,(1) but certain fundamental eras are virtually blank. The most significant gap is the period between the two World Wars, an age in which perceived excesses in recreational drug use were met by a vigorous international suppression effort centered in the League of Nations.(2) This was the crucial time when steps were taken to more narrowly define legitimate drug consumers and to discipline legitimate drug firms, pharmaceutical traders and dealers, into servicing only this shrinking market. By convincing the large firms, through one means or another such as profit stabilization through cartel pricing, to hold production and to tighten their marketing procedures, League reformers pushed manufacturing for the illict market ever more irrevocably into the hands of organized criminal entrepreneurs. The ranks of the latter were swelled by individuals from small firms driven to the edge by state regulation and the inhibiting power of the cartel.

At the beginning of this period, the illicit trade in refined narcotics depended to a large extent on diverting legally manufactured narcotics. Organized criminals were typically located at the tag end of the drug-manufacturing and marketing process. In combination with manufacturers or various and numerous middlemen in transportation and retail outlets, they diverted a portion of the product to supply de-certified consumers. Before the two decades were over, however, professional criminals were almost alone at the beginning of the process owning clandestine drug factories around the world. Their opportunity was a consequence of the League's limited success in controlling through exhortation and regulation legitimate production.

This is not to suggest that European organized criminals ruled the international drug trade by the start of World War II. They had plenty of competition especially in the Far East, where huge Chinese, Japanese and Korean syndicates were formidable powers. It is rather to argue that European criminals played crucial roles in the years of greatest structural change in the production and trafficking of narcotics. When the very large European firms stopped producing for the illicit market, criminal firms, several already in existence having carved out a meager if not scrawny portion of the illicit trade prior to World War I, assumed a much more important position.

The model to consider is the American experience with alcohol prohibition. As legitimate American producers were forced out of business, underworld syndicates began to manufacture what they only used to sell and distribute. The most successful were able to vertically integrate their alcohol interests--producing, transporting and marketing whiskey or beer. This was so in many but not all cases. Some of the original manufacturers were able to stay in the market by striking deals with organized criminals who would "front" for the producers and market their product. Foreign manufacturers in Canada, Ireland, the United Kingdom, etc., neither needed nor wanted help in production. They simply serviced the totally illicit market by selling to criminal wholesalers.

The international traffic in drugs, like alcohol trafficking, drew on the talents of numerous groups. In the growing drug underworlds of Europe, the Middle East, and even the Far East, Jews and Greeks stood out as prime participants, among the European master craftsmen of contraband trading. Their participation during this crucial period is an important part of their social history generally overlooked and certainly unappreciated. Organized criminal activities tend to integrate ethnic and national minorities within often powerful coalitions or syndicates which have many significant points of contact with legitimate society. Understanding any aspect of the social world of organized crime can only enhance social history in general and lead to a more profound realization of the social experience of the various people involved.

THE LEAGUE'S DRUG POLICIES

International efforts to stem world drug usage reappeared after a hiatus brought about by the first World War. Under the banner of the League of Nations the signatory nations created the Advisory Committee on Traffic in Opium and Other Dangerous Drugs to coordinate the struggle. Additionally, ratification of the Versailles Treaty brought with it "de jure" ratification of the Hague Convention of 1912. One of several early international conferences attempting to cap world drug consumption, that meeting "provided for the gradual suppression of opium smoking, limited the use of manufactured narcotics, the opiates, and cocaine to medical and legitimate purposes, and subjected their manufacture, trade and use to a system of permits and records."(3) The League's initial strategy--controlling drug consumption by regulating the production of the international pharmaceutical industry--was thus in place at the start. Profiteering in morphine, heroin, cocaine, etc., was a consequence of nationally unregulated pharmaceutical plants and their relationships with smugglers from Europe's organized underworlds. The League contended with both elements, although it usually left implicit the most important and interesting connections between industry and organized crime.

Some nations believed the control of illicit drugs could only occur by abolishing the private ownership of drug factories. Several years into the policy of control, for instance, the government of Estonia wrote the League's Advisory Committee in April 1928 that the illicit drug traffic continued "on an enormous scale," backed by giant financial resources. It appeared impossible to restrict the amount exported from the producing nations of "raw opium and coca leaves" to the much smaller quantity necessary for the world's medical needs. Thus Estonia recommended that "all Governments Members of the League and parties to the Opium Conventions" should takeover the factories manufacturing dangerous drugs.(4) That was a telling admission but hardly a practicable policy. Among its many drawbacks was the elementary one that several important League members were colonial powers and ran overseas economies which were dependent, to one degree or another, on large-scale drug production and trading. The British, Dutch, French and Portugese come readily to mind. There appeared this revealing inconsistency: within Europe the major villains were private manufacturers and the international underworld; in colonial areas, on the contrary, it was government policy to encourage narcotic production and consumption and the villains were unwanted temperance efforts.

Putting aside the colonial incongruities which were impossible to deal with in this setting, the League worked to encourage government regulation of drug manufacturing to curtail production. It is also apparent that the League concentrated on manufacturing because it had so little control over the world's raw opium harvest. Four of the six or seven great opium-producing countries in the world, namely Turkey, Persia, China and Russia, had never agreed to either the Hague Convention or to any other similar international plan.(5)

Meanwhile, the regulatory issues, even though affecting only part of the drug-producing and manufacturing world, went on apace. Whatever nations participated, regulation first and foremost was dependent on the resolution of formidable data-gathering problems. The Advisory Committee had to determine what production level was consistent with "legitimate medical" needs, thus theoretically enabling it to identify that portion of morphine, heroin, and cocaine which catered to recreational users and medically "de-legitimized" addicts. It then had to convince nations to limit manufactures to satisfy just legitimate needs.

The first advances in setting the necessary administrative apparatus to deal with proper drug documentation occurred during the League's Geneva Conference of 1924-25. The Convention agreed to in 1925 (which became operational at the end of September 1928) established a Permanent Control Board to survey drug factories, to collect and analyze national quarterly reports on imports and exports of proscribed drugs, and to conduct other similar surveying and statistical tasks. Each assignment presented unique difficulties. One of the most intractable was that of identifying factories authorized to manufacture narcotics.

The League asked governments to catalogue their factories.(6) Using their replies, the League's Secretariat produced a list of countries claiming no factories, and another with a detailed reckoning of manufacturers. But a glance at the lists indicates the daunting survey problems associated with regulation and the many potential entry points for "leakage" into the illicit traffic. It was quickly evident that nations were uncertain how to list pharmacies and secondary manufacturers of compounds such as extract and tincture of opium, opium powder, tincture of coca, phials of cocaine and morphine, laudanum, and so on.(7) Most concluded the League's intention was a list of their leading, primary manufacturers, not one including all the derivative operations though it was manifest that some quantity of the illicit trade spun-off from these secondary firms.

The major European producers of refined drugs were Germany, France, and Switzerland. Great Britain and Italy manufactured on a smaller scale, as did The Netherlands which had a cocaine factory in Amsterdam and another turning out opium alkaloids in Apeldoorn. The following were the primary factories in the three predominant nations:

Figure 1. DRUG FACTORIES IN GERMANY, FRANCE, SWITZERLAND, GERMANY

Factory	Place	Drugs
C.H. Boehringer Sohn, A.G.	Hamburg	morphine, heroin, cocaine
C.F. Boehringer & Sohne, G.m.b.H.	Mannheim-Waldof	morphine, heroin, cocaine
Hoffman, La Roche, A.G.	Grenzach (Baden)	morphine, heroin, cocaine
Gehe & Co.	Dresden	morphine, heroin, cocaine
Knoll A.G.	Ludwigshafen a/Rh	morphine, heroin, cocaine, dicodide, dilaudide
E. Merck	Darmstadt	morphine, heroin, cocaine, eucodal
Chininfabrik Braunschweig-Buchler & Co.	Braunschweig	cocaine
I.G. Farbenindustrie A.G.	Leverkusen (Rheinland)	heroin

FRANCE

Comptoir français des alcaloides	Paris	morphine
Societé industrièlle de Chimie organique	Paris	morphine
Societé de produits chimiques de l'Quest	Paris	morphine
Societé d'industries chimiques de l'Quest	Paris	morphine
Laboratoire Clin, Comar et Cie	Paris	morphine

(Figure 1 continued)

Factory	Place	Drugs
Etablissements Roques et Cie	Paris	cocaine
Roessler, Fils & Cie	Mulhouse	morphine

SWITZERLAND

Factory	Place	Drugs
Dr. Hefti, Fabrique de produits chimiques	Altstetten	morphine, heroin, cocaine
Sandoz (Fabriques de produits chimiques)	Basle	morphine, heroin,
F. Hoffman-La Roche & Cie. S.A.	Basle	morphine, heroin, cocaine
Soc. de Chimie industrièlle (Ciba)	Basle	morphine, heroin, cocaine
Th. Muhlethaler S.A.	Nyon	morphine, heroin

With production figures for the individual factories unavailable, it is not possible to say more than that Paris was the most significant European drug manufacturing city followed by Basle. Nonetheless, the licensed factories (and pharmacies) were identified and some control over the system in Europe began to be exerted. This was partly so because cooperating nations had agreed to a procedure for licensing imports and exports using government authorized certificates. Export certificates, for instance, were supposed to be issued only upon production of proper import certificates from importing countries, and vice versa. It was actually a tracking program in which mandatory import and export "certificates" could be matched. Tracking of the legitimate drug trade, it was believed, would reveal what drugs from which producers moved across the line into "underground" markets.(8)

Information from quarterly reports and scrutiny of import and export certificates enabled the League to discover what it called "discrepancies." The way it worked was this: if country A reported x amount of heroin exported to country B, but country B could only certify imports of y amount, the difference was a "discrepancy." There were both positive and negative discrepancies depending on whether acknowledged imports were greater or less than corresponding exports. Large discrepancies indicated in many cases a diversion of stock into an illicit market.(9) Confirmation of this came from many quarters and was duly reported in sections of the League's 1931 report summarizing and analyzing the international traffic from 1925 through 1929.

In 1928, as an example, France exported 346 kilos of morphine to the United States which the U.S. Federal Narcotics Control Board reported were "illicitly imported." There simply were no import certificates in this matter. The U.S. noted the French, when asked about this, offered no explanation.(10) Data from 1927 showed Germany had received 440 kilos of morphine from France which were never cleared with the Germany Ministry of Health. This mystery, Germany contended, could only be solved if the French provided details of their exports and consignees in Germany.(11) Similar reports on French exports came from Belgium, Denmark, and Greece. And The Netherlands reported confiscating 62 kilos of French morphine in Tandjong Priok (East Indies) in 1928.(12)

France was not the sole European manufacturing country involved in supplying drug racketeers. Germany exported 159 kilos of morphine to Estonia in 1925 which could not be accounted for and was presumed to have fallen into the hands of criminals.(13) The Swiss sent almost one thousand kilos to France which failed to show up in the corresponding French tally for 1928. There were much larger differences in earlier transactions between Switzerland and Japan. In 1925 the Swiss pharmaceutical giant Sandoz exported over 1,300 kilos of morphine to a Japanese firm which had no record of the transaction. The League believed the consignment had passed into the illict international traffic.(14) Keeping track of discrepancies in the import/export trade was very difficult; analyzing them was trying and diplomatically touchy.

Another part of the League's control plan which was especially irksome was the call for national drug "quotas." This evoked the inevitable tension between devotees of the "free market" and those wishing to inhibit the traffic. The inhibitors saw themselves, by the way, as the "pioneers" of a worldwide planned economy. International drug control was the first concerted attempt made by the Governments of the world to regulate a branch of a single industry internationally, from the point the raw materials entered the factory to when they finally reached the legitimate consumer. This effort embodied the principles of a planned economy commanded by "a real international administration."(15)

Limiting narcotic consumption meant establishing production allotments--"The fraction . . . to be manufactured by each of the manufacturing countries to be fixed in advance by means of agreements concluded between them according to a system of quotas"--and these were a source of contention. Japan (which was a major narcotic power acting primarily but by no means exclusively in China and Southeast Asia) insisted on the "free trade principle" aiming to allow all nations, "including those only manufacturing for their own requirements, to share in the export trade in narcotics in the event of their receiving legitimate orders from abroad."(16)

Beyond Japan's protests were Turkey's complaints. Turkey had spent the better part of the nineteenth century (up to 1914 in fact) losing huge chunks of its vast European territories. It was reduced in Europe after World War I until all that was left was a bit of the peninsula upon which Constantinople (later Istanbul) rested, and also temporarily lost ground in Asia Minor while Greece controlled the important city of Smyrna and its adjacent territory. As Turkey's territory and national economy dwindled, the relative significance of opium production naturally increased.

Turkey had never signed the Conventions governing the drug trade, but in 1931 declared it was considering doing so if certain problems were overcome. At that time Turkey headed "the opium producing countries of Europe and alone practically supplie[d] the European markets."(17) Turkey's concern over quotas stemmed from its decision to manufacture drugs from its own superior opium whose morphine content was reportedly a potent 14 percent. In 1931 Turkey announced it had three drug factories at Istanbul which had already manufactured 2,300 kilos of morphine and 4,300 of heroin in their first six months of operation. Moreover, the Turkish government acknowledged past illegal drug trading but stated that was no reason for the League now to exclude Turkey from the licit manufacturing market. The fact was Turkey "has the right to demand the allocation of a [manufacturing] quota on the same footing as other countries."(18) And if Turkey's "just" demands were not met then it would continue on its unregulated way--Turkey "is prepared to accede to the Geneva Convention and to limit the manufacture of narcotic drugs, provided that it is given a fair quota."(19)

Joining Turkey in protest was Yugoslavia which believed the illicit trade would be greatly increased unless "all civilized countries" accepted the Geneva Convention on limiting manufactures. Failure would mean that producing countries would inevitably become manufacturers searching for new markets. Additionally, Yugoslavia claimed to be suffering because of its adherence to League policies and the pricing policies of the growing cartel of drug manufacturers. The cartel's impact upon the trade was obvious. By January 1930 the price of raw opium had fallen almost 60 percent and future sales were uncertain; however, the prices of the manufactured drugs remained stable. Finally, Yugosalvia noted that "if Turkey is authorized to manufacture, an equitable quota should be allocated to Yugoslavia also."(20)

Turkey, Yugoslavia and Greece were the only opium growing nations in Europe. Greece was a small producer with the capability of exporting approximately 25 tons (almost 22,800 kilos). Yugoslavia's production, while not nearly that of Turkey's, was considerably greater than Greece's. The Yugoslav government provided a comparison between its opium production and Turkey's which indicated, among other things, that Turkey's admitted 1928 export of 345 tons (313,248 kilos) still left around 50,000 kilos for either home consumption or unidentified markets.

Figure 2. OPIUM PRODUCTION IN TONS (21)		
	Yugoslavia	Turkey
1925	148	240
1926	100	360
1927	55	280
1928	205	400
1929	38	200
1930	150	480

Although it is not clear to whom Yugoslavia exported opium, Turkey's customers (for at least the year 1928) are known and they show Turkey's key role in European opium supply.

Figure 3. 1928 OPIUM EXPORTS FROM TURKEY (KILOS)(22)	
Imported By	Amount
France	228,632
Germany	20,632
The Netherlands	12,409
Great Britain	11,452
U.S.A.	8,021
Italy	7,422
Greece	5,898
Japan	4,582
Syria	2,418
Belgium	1,520
Egypt	1,373
Soviet Republics	763
Spain	230
China	228
Miscellaneous	7,668

France provided an extraordinary market for Turkey's opium. In 1928 France imported 251.5 tons of Turkish opium, almost 3 times the amount reportedly imported by the rest of the world. This figure confirmed that France was Europe's major producer of opium derivatives and thus the likeliest center for diverting narcotics into the underground economy. "It was during the 1920s," one drug expert has written, "that Paris emerged as the operational center for the illicit procurement of legally produced heroin, morphine, and cocaine . . . ."(23) This was the reason why many other nations when reporting to the League's drug inspectors so tersely noted large irregular imports of French narcotics and acidly remarked the French government offered neither an explanation nor clarification. In one case involving the seizure of cocaine and opium on a French ship in a U.S. port in 1928, the U.S. unsuccessfully asked the French for information for more than two years.(24)

THE REALM OF PROFESSIONAL CRIME

The League's compilation of worldwide drug seizures reveals the striking French involvement. But France was not alone; sixty nations and colonies reported illicit transactions and seizures in 1929. In detailing the hundreds of reported criminal events, some of the methods of smugglers were enumerated: (1) using counterfeit and stolen import/export certificates; (2) establishing phony chemical and pharmaceutical companies; (3) working with corrupt insiders in legitimate factories; (4) employing false bills of lading; (5) hiding contraband in ingenious ways; (6) shipping drugs to "unregulated" colonies, territories and certain intermediary nations whose function was a flourishing re-export trade; and (7) organizing clandestine factories. This last, of course, was increasingly important as time went on.(25)

By the latter 1920s, the League and several national police forces were searching for better methods of controlling the illicit traffic. Over the course of the decade as legislation had criminalized recreational consumption and severely restricted retail distribution by physicians and pharmacists, professional criminals had more conspicuous and important marketing opportunities. So pressing did underworld activity appear, the League's Advisory Committee decided to create a "Black List" of criminals. A draft Convention was prepared. However, the draft produced several cautions during the 34th meeting of the Advisory Committee's Fourteenth Session on 2 February 1931. Swiss and French representatives expressed most concern principally over controlling such a list's circulation and whether company names should be included. In response it was noted the Black List was not intended for publication but for national policing authorities. The necessity of including companies was defended using an example of a large Swiss firm which before 1928 was considered completely legitimate, but then proved quite the opposite. At this meeting it was decided that similar questions over the draft should be submitted to a Sub-Committee.

Black Lists were prepared and annually updated. They are difficult to locate, however, and researchers must use those portions that were inadvertently made available from time to time. What appears to be an update of the 1936 list (published along with other reports from the Advisory Committee) named 37 racketeers with diverse national backgrounds: Japan, Italy, the U.S., Switzerland, Austria, Lithuania, Germany, Greece, Yugoslavia, France, Poland, and Bulgaria. The largest cluster represented Turkey and those European parts of the former Ottoman Empire which had shaken loose during the past century--Yugoslavia, Greece, and Bulgaria. These traders who serviced an assortment of markets were primarily supplied by manufacturers in Turkey, France, and Bulgaria.

Some of these dealers such as Metodi Lazoff trafficked between Bulgaria and Czechoslovakia while others like David Levy moved drugs between Turkey, the Near East, France and the U.S. There was a flourishing trade between Yugoslavia, Bulgaria, and France. Traffickers Nissim Calderon and Mikhaili Anapniotis who were headquartered in Turkey smuggled into Egypt and Ethiopia. This complicated world of Near Eastern drug smuggling, involved several sets of smugglers moving stock back and forth across the Mediterannean (from Lebanon to Marseille, for example) and up and down the Red Sea and Suez Canal destined for Cairo and Palestine.(26)

Ethnicity and the Drug Trade

There were a disportionate number of Jews on the League's Black List of 1936. (Some of these racketeers had close ties to New York's narcotic markets then dominated by American Jews like Yasha Katzenberg, Jake Lvovsky, Mendy Weiss, the Newman brothers, Sol Gelb, etc.).(27) The 1936 list was not an ethnic or national aberration. Throughout the interwar period European Jews (for that matter, Middle Eastern ones also) were disproportionately absorbed in drug smuggling. This is partly explained by familiar factors. Their long exclusion from European civil society forced them into the most extreme social and economic statuses. Pushed into borderline commercial ventures by the vagaries of Jewish existence in an alien, hostile world, dealing in contraband was inevitable. However, smuggling and illicit trading were not exclusively based on social isolation and economic dislocation. There are other historical causes to note. There were times and places in nineteenth century Europe when Jewish marginality led to the development of skills and attitudes necessary for "modern conditions" which appeared to remove any need for illicit talents. Marginality in those cases was initially advantageous although in time the advantage turned. A brief discussion on Romania will make this clear.

Romania had lived for centuries under foreign control.(28) Some sections of what became modern Romania were within the Hapsburg Empire, others were "subject states of Turkey and later, of Russia." All told the region was dominated by "foreign masters" and routinely decimated by large armies. Around the turn of the eighteenth century, the region's diverse principalities were controlled by representatives of foreign powers--Russia and Austria first, then Britain, France, and Prussia.(29) Foreign traders particularly Jews, Armenians, Greeks, and Bulgarians, were persuaded by offers of physical protection and economic advantage to settle in Romania.

By 1830 many of the trade restrictions which had propped up the antiquated guild systems were dumped, and the foreign trader "relieved of certain duties and taxes, supported by his consul, benefiting in many instances from a more enterprising mentality and more up-to-date methods, enjoyed such advantages that some Romanians sought foreign nationality as a key to commercial success."(30) Romanians were the rural masses sunk in poverty and exploitation, made worse by the impact of "freer trade." The Romanian middle class was primarily represented by Jewish immigrants and the other foreign entrepreneurs. But when nineteenth-century nationalism finally asserted itself, Romanians "had to affirm themselves" against Turks, Greeks, Russians, Jews, Armenians, Bulgarians, and so on. The newly-awakened "volkish" sentiment turned against the foreign entrepreneurs, particularly those operating under the protective system. Romanian anger became predominately focused on Jews in Moldavia and Greeks in Wallachia. At the turn of the twentieth century, Jews were being reviled "as usurers and tavern-keepers."(31)

Peasant anger was directed against Jews who personified cultural modernism and industrial development. Most industrial investment funds came from foreign capital which owned over half of Romania's oil industry. Banks and insurance firms were also dominated and controlled by foreign capital. Instead of focusing on Romania's position as a "colonial" outpost under the sway of British, German, and other outside capital, more and more attention was turned on the foreign entrepreneurs, in particular Jews who had arrived under the liberating and protecting stimulus of foreign interests. They embodied the mystifying and dangerous modern world; their children dominated the key professions of engineering, law, finance, medicine and pharmacy. On the stock exchange in the capital city of Bucharest almost every broker was Jewish. In the province of Bessarabia, historian Eugen Weber finds that Jews ran virtually every drug store.(32) In the interwar years the social distance separating classes and people in Romania was increasingly resented: the "poverty of the masses was emphasized by the wealth of the few, the luxury of the upper classes, with their gambling clubs, their American cars, and their Paris fashions."(33) The envy produced was explosive and concluded in movements characterized by violent chauvinism, xenophobia, and leader cults.(34)

Whether at the worst or best of times, Jews and other minority groups in Romania (indeed throughout central and eastern Europe) were familiar with organizing crime. The unsuccessful and persecuted did so out of immediate necessity, the successful naturally waited until their advantages called forth the familiar accusations of impropriety and their civil standing eroded.

Greeks were also disproportionately involved in narcotics, drawn to smuggling for particular historical reasons. Ottoman Greeks provide an important example. Beginning in the late sixteenth century, the vigorous control the Ottoman Empire had asserted over internal trade was modified in the face of the rise in grain prices and raw materials in Western Europe. Ottoman merchants were lured to more lucrative markets outside the confines of state control and "contraband trade carried the day."(35) Much of this illicit commerce was carried out in coastal areas by Greek ships harbored on Aegean islands. The very ubiquitousness of smuggling meant significant change had already begun concerning the Empire's economic position within Europe. The weakening of political control over the traditional methods of internal trade signalled a "disarticulation," or unravelling, of the system. This included the overthrow of worn out modes of agricultural production as capitalism commercialized the countryside. The fortunes of certain ethnic groups within the Empire were seriously altered by these changes, sometimes for the better. Greek dominance of illegal commerce, for instance, led to the rise of a powerful Greek bourgeoisie.(36)

Jews and Greeks were thus well placed to participate in the illicit drug trade. Jews in Romania held strategic positions in medicine and pharmacy which were key entry points for drugs moving into illicit markets. The Greeks' long experience in contraband trading and their location between opium growing Turkey and Yugoslavia were both significant. Additionally and perhaps most importantly, Greeks and Jews were a dispersed, talented, commercially-minded peoples. There were Greek and Jewish communities around the world in the most far-flung cities linked one to another through kinship and "village solidarity." The material for international criminal syndicates was there, needing only the right circumstances such as economic dislocation or tempting opportunity to form.

European drug syndicates were constructed by individuals able to tap into an international community taking advantage of ethnic and kinship ties though never confined by them. Syndicates grew, changed, diminished, and grew again as circumstances warranted. They are sociologically notable for their adaptability, the semi-independence of many members, the ephemeral nature of gangs, and for their unique blending of nationalities and ethnic minorities.(37) The Zakarian Organization based in Cairo, Egypt, is a case in point.

Zakarian was an Armenian with a Greek partner who turned to the drug trade in 1925. They opened a carpet shop as a front and quickly established themselves as large wholesale heroin dealers. In the beginning they purchased drugs on the flourishing narcotic market in the ancient port city of Alexandria. A commercial crossroads for several thousand years, Alexandria's importers and wholesalers were Jews, Greeks, Armenians, Egyptians, Turks, Lebanese, etc. Soon, however, Zakarian moved to European suppliers who promised cheaper rates and "a greater regularity in consignments."(38)

Zakarian went to Vienna in 1926 and teamed up with Ludwig Auer, a chemist who had already been fined for importing heroin from The Netherlands without a permit. Auer supplied him with "Morphium Benzoylicum" obtained from the French manufacturing firm Roessler in Mulhouse. Auer was paid a commission by both Roessler and Zakarian. A few years later this arrangement ended and was replaced by a new Zakarian-Auer connection. This time Zakarian used a pharmacist in "Telaviv, Palestine," as his intermediary. After a few shipments this was disbanded when Zarkanian found another more lucrative source of supply. "Two young Polish Jews [brothers] by name Zelinger living in Vienna," had a better proposition. They promised to deliver as much heroin as Zarkarian could handle directly to him in Cairo. He was to remit payment through a Vienna bank. The Zelinger's source was the Swiss drug factory owned by Dr. Hefti and located in Alstetten near Zurich.

Zarkarian and his partner were just some of the Zelinger's customers. The brothers had their own syndicate based in Vienna which contained at least seven others. This group included a "leatherware merchant" who built "cabin trunks" with secret partitions for smuggling heroin, an accountant, a commercial traveller, a druggist, and the manager of a Viennese pharmacy. The Zelinger syndicate was Jewish with the exception of the accountant who was a Protestant.(39) Vienna harbored other Jewish drug merchants whose activities mixed and mingled with those already mentioned. One complementary Viennese syndicate of greater scope was composed of Russian, Polish and Palestinian Jews who sold to dealers in America, the Far East and Egypt.(40)

The Turkish manufacturing scene was composed of a rich blend of ethnic and national groups. Two of the three factories acknowledged by the government to operate under official authorization were interestingly staffed. The Orient Products Co. in Istanbul's Taxim Quarter began manufacturing operations in 1927 under the direction of two Japanese brothers. They bought their "concession" [license] from the government and later worked under a permit issued by the Ministry of Public Health. Another licensed group was an important French manufacturing firm which formed the Djevdet Bey Factory producing morphine, cocaine, and heroin. The factory's Board of Directors was composed of "2 Armenians, 1 Turk and 2 French men."(41) Both of these factories--Orient Products and Djevdet Bey--supplied the booming illicit market in Alexandria. By the end of the 1920s, around 50 kilos of heroin per week (over two and a half tons a year) were smuggled ashore at Egypt's great port by the crew and staff of steamship lines.(42)

Other drug syndicates, like the Board at Djevdet Bey, were a hodge-podge of nationalities. The Swiss-based "Basle Organization" had nationals from several European countries. This syndicate, like Zakarian's, was smuggling heroin into Egypt. A primary smuggler in this group was a Romanian Jew, Maurice Grunberg, who lived in Cairo where he ran several businesses including a fair-sized printing establishment.(43) Grunberg's Swiss connection was Dr. Fritz Muller, a German subject holding a permit from Swiss Authorities for the manufacture of drugs. He had his own laboratory in Basle.(44)

Dr. Muller was like the hub of a wheel with spokes radiating out. One went to Grunberg, another (along which 350 kilos of heroin travelled) led to a Greek dealer, a third to an importer/exporter in Istanbul who sent heroin on to Alexandria, and a fourth proceeded to two "French citizens of

Mulhausen and Strasbourg."(45) There were still other spokes including a chemist in Zurich supplying heroin for Italian consumers and smugglers who moved it to Alexandria and Cairo through Genoa and Trieste. The Basle Organization, really a shifting coalition of Greeks, Italians, Swiss, Germans, French, and Romanian smugglers, was an important mid-European channel" for Swiss heroin smuggled to the Near East.

Another prominent European-based drug syndicate to surface during the interwar period started in 1927 when Elie Eliopolus, a Greek (also allegedly a Jew) living in Paris, found a South American money-man willing to finance a drug partnership. Their scheme was to buy French manufactured narcotics and market them in the Far East. Eliopolus had a contact named David Gourevides (aka Gourevitch), a Russian Jew quite familiar with the drug scene in Northern China. Gourevides had the best European connection in the Far East. This was Jean Voyatzis, a smuggler for over 25 years with a flair for organization, providing leadership for dozens of Greek dealers located in China. Europe had been supplying heroin to China through the ports of Tientsin, Shanghai and Dairen for quite some time.(46) Eliopolus and Gourevides travelled to Tientsin in November, 1927, so that proper introductions between the principals could take place.

The Eliopolus association was taking shape with its Far Eastern wing secured. It next turned to police protection and acquiring narcotics. From 1928 through 1930, Eliopolus was under the protection of a Mr. Martin called "Zani" of the Prefecture de Police. The deal with Zani was made on the understanding that Eliopolus would not sell in France. For his part Zani received information on other traffickers and most importantly 10,000 francs every month.(47) Eliopolus and his partners purchased morphine from two Parisian firms--Comptoir Francais Des Alcaloides and Societe Industrielle De Chimie Organique (called SICO). The former factory was run by a man named Kieffer who also had a busy cafe on the Rue Royale, and Paul Mechelaere known as a "notorious Belgian trafficker."(48) SICO was owned by an individual named Devineau. With drugs regularly at hand, Eliopolus developed markets other than the Far East. He worked with August Del Gracio an important American smuggler who purchased drugs for one of America's largest organized crime gangs. There were still other channels including a Berlin one which was handled by an individual known as "Seya Moses."

The Tide Turns

By the end of the 1920s, the European drug scene was changing. France had finally reacted to its notorious position as an illicit drug trader and curbed opium imports which in 1929 fell by around 2/3rds, down to 80,840 kilos. French authorities also admitted "that there had been a leakage which had not been prevented by the regulations in force." In this instance they were referring to drugs shipped to Algeria and other colonies. French internal controls were subsequently strengthened by a special Decree on 20 March 1930 which finally required export authorization for its colonial trade. More importantly, they closed three drug factories in Paris and withdrew "the trading licenses of all firms mentioned as having been involved in important cases of illicit traffic."(49) This was certainly a critical period for criminal entrepreneurs marking the time when licit European manufacturing began to significantly dry up and hence when criminal firms turned an important competitive corner. The immediate impact on French drug manufacturers was threefold: some were driven out of business; others burrowed deep into the Parisian underworld; and still others left town moving their operations to Istanbul and forming new partnerships. Devineau from SICO and Mechelaere from Comptoir Alcaloides founded the Kuskundjuk factory in Istanbul.

The new Istanbul manufacturers had an array of customers representing various transcontinental networks. Eliopolus, who now purchased all his drugs from these Turkish factories, was not alone. Del Gracio, one of SICO's most important customers before the French cleanup, helped Devineau and Mechaelere relocate in Istanbul by either giving or lending them $75,000.(50) He remained a valued customer to say the least. The Kuskundjuk operation marketed through several channels including a freight forwarding agency headquartered in Prague, Czechoslovakia, with branches in Istanbul and Hamburg. The entry into the Hamburg underworld was smoothed by Karl Frank, a notorious smuggler who also transferred drugs for Eliopolus.

The move to Istanbul aroused difficulties as well as solving problems. The treacherous social relations among professional criminals were exacerbated by the shift from Paris to Istanbul. Drug barons cheated one another with absolute abandon. Manufacturers tried to swindle the buyers, while the buyers in turn threatened to murder them. And everyone seemed dedicated to informing. Relations among the traffickers in Instanbul were so unstable that one gang, which had its own factory, turned from manufacturing to informing for a share of the government's rewards on seized contraband.(51) It was only a small step to outright blackmail of the other factories. The gang's stock of drugs grew through extortion of the other refiners. The muscle for this syndicate came from Paul Carbone, a Corsican-born racketeer once expelled from Egypt on "white-slavery charges." Carbone is credited with being one of the founders of the Marseilles heroin gangs and establishing Marseilles' first heroin conversion lab.(52) The origins of the infamous "French Connection of the 1960s" are to be found in the changes and personnel working the Istanbul underworld in the interwar period.

The rise in gang conflict made Istanbul a somewhat less desirable locale around 1932 or so. It was at this point also that the police in Egypt, Germany, and Austria arrested several of the more important Istanbul-based traffickers. Given these conditions, some observers felt prominent traffickers might leave Turkey for a spell. It was believed that Bulgaria was among the likely destinations as traffickers could buy protection from a nephew of the Minister of War.(53) Bulgaria also had one known illicit drug factory located at Radomir, a short distance southwest of Sofia.(54) Bulgaria, however, couldn't begin to offer the criminal opportunities of the Far East.(55)

When it came to narcotics, the action in Asia was so expansive that it seemed a world apart from Europe and the Near East.(56) Part of the dynamic Asian trade was in the hands of European smugglers. Whether importing European manufactured drugs into China during an early period, or exporting Asian heroin and morphine into the U.S. and Europe later, there had been groups of European smugglers in Chinese cities for decades. They formed segments of underworld communities which expanded in the early 1930s as more European racketeers migrated east from Istanbul. The allure of the rapidly expanding Chinese trade coincided with the temporary dislocations experienced in Turkey.

EUROPEAN SMUGGLERS IN THE FAR EAST

Through a series of calamities China had become the world's greatest opium-producing and consuming nation. Without wishing to belabor the point, opium trading which was first inflicted by the British on China, "fell into the hands of the Chinese underworld after the turn of the century." Historian Sterling Seagrave writes that "during the warlord era that followed, opium was also the main source of revenue for the military rulers who controlled China province by province; taxes on its cultivation and transportation, opium dens and paraphenalia provided the sums to meet both military and civilian needs of these pocket dictatorships."(57)

In the 1920s and 1930s, when China was wracked by civil war and then war with Japan, the most reliable currency for those contending for power was opium. It provided racketeer Chiang Kai-shek, leader of the "Nationalist" forces, with his primary source of revenue. Chiang belonged to one of China's largest organized crime conglomerates known as the Green Gang; a vast underworld society which had virtual sway over the broad branch of the opium traffic travelling on China's great Yangtze River ending at Shanghai's delta region. Shanghai was "founded on a great brown swamp of opium tar." Opium granted Chiang "a source of giant revenues that guaranteed him great amounts of secure foreign currency."(58)

Much of China seemed like a measureless opium farm, although there were regional exceptions such as the province of Shansi.(59) But China had not been traditionally a major manufacturing area for morphine and heroin. Even though there had been some small refining in the 1920s, China at that time was still chiefly a market for refined narcotics filled by European and increasingly Japanese traders. One historian has noted (perhaps with some exaggeration particularly as no source is given) that China imported enough heroin from Japan in 1924 to enable the entire population of 400,000,000 to have four hearty doses a year.(60) Whether the figures are accurate or not, they do indicate Japan's massive impact on the China market of the 1920s. European traffickers still remained on the scene though clearly their place as China's prime providers of heroin and morphine had gone to Japan. A few years later, the Japanese faced stiff competition from Chinese producers themselves.

Seagrave comments: "The Chinese were blessed by a constant supply of the very finest heroin, thanks to Tu Yueh-sheng [the leader of the Green Gang]. It was customary in the streets of any Chinese city simply to buy and swallow pills of relatively pure heroin or sometimes to smoke it in pipes in the form of pink tablets."(61) The Green Gang produced so much heroin that it became an important exporter to France and the United States in the 1930s. Chinese heroin travelled with official approval through the French Concession (a territory under the civil and police administration of France) in Shanghai. Corruption in the French Concession was total; most of the police detectives were members of the Green Gang; the top police officials were completely under Tu's influence.(62)

Shanghai was the most important crime city in China and the European smugglers, as one knowledgeable observer remarked, shifted restlessly between it, Tientsin, and Dairen, two strategically important port cities. Tientsin was Peking's port, and Dairen was a main gateway to Manchuria. Dairen nestled at the end of a peninsula to the east of which was Korea which had been annexed by Japan in 1910. A rail line ran from Dairen north through Mukden and Harbin (two other cities with important European smugglers) and then turned west running through Manchuria and into the Soviet Union.(63)

The European drug smugglers in China represented a number of ethnic and national groups whose criminal activities spanned several collateral illicit enterprises. They spun intricate webs in which legitimate occupations masked illicit ones, and they typically operated from the privileged colonial sanctuaries found in Chinese commercial and trading cities. They dealt with Japanese and Chinese military and political officials, and organized crime syndicates representing Chinese, Japanese, or Korean interests.

A survey of European smugglers in China in 1933/34 was requested by the three-year old U.S. Federal Bureau of Narcotics which wanted information on all aspects of the Chinese drug scene. Carried out by American Treasury officers the survey of European smugglers turned up approximately 179 individuals representing a mixed lot of Germans, Armenians, Russians (including Tartars and Georgians), Central and Eastern European Jews, Greeks, Serbs, Poles, Austrians, Bulgarians, Czechs and Italians. The figure is not more exact because a few were likely counted twice, once under their real name and again under an alias. In spite of these difficulties, the survey has more than enough convincing detail to demonstrate its soundness.

A little over 40 percent of the surveyed racketeers were Greeks. The next sizable group was Russian but it could be counted in several ways--those identified only as Russians made-up 27 percent of the total; adding Russian Jews and Russian Tartars to the figure, however, raises the Russian complement to 35 percent. Identifying Jews, whether Russian, Polish, German, Austrian, etc., was thought important. They totalled about 10 percent of those counted.(64)

Smugglers were mobile and opportunistic, trafficked in several types of contraband, rapidly entered into and quit partnerships, and seemed almost as likely to forge associations across ethnic and national lines as along them, although kinship was important. In many instances, the individuals and their circumstances defy any easy or obvious classification. The adventures of A. J. Avramow, a Bulgarian working in Shanghai for a firm which manufactured bank notes were commonplace. He was also in the drug smuggling business but desired to become a manufacturer. Consequently, Avramow invested $45,000 for a heroin factory in Peking. The deal failed when his partners, a "Serb" from Yugoslavia and a Japanese, cheated him. Following this flop, he tried diamond smuggling hoping to use a diplomatic post as cover. Avramow angled to become the local Bulgarian Charge d'Affaire.(65) In this he was no doubt inspired by his former Serbian partner, E. Bechich, who had been in the Serbian diplomatic service. Bechich was out of the foreign service when he joined with others in the "Jugo-Slavian Pharmacy." Employing the pharmacy to illegally import European narcotics and perfumes, Bechich became the leader of a Yugoslavian gang dispensing both commodities.(66)

China's conflicts guaranteed that several drug traffickers would be in the munitions racket and that others would be foreign agents. E. W. Eickhoff, a German in Shanghai, was a notorious dealer in arms and ammunition, as was Gustav Hoeger a British subject of German origin who had lived in Russia. In Shanghai, Hoeger represented a Canadian weapons firm which was busy smuggling guns into China. Illicit arms dealers in Mukden met at the Hotel Keining whose owner, F. Keining, was known as an unofficial representative of Germany's E. Merck drug factory in Darmstadt.(67) In Tientsin, S. L. Bonetti, an Italian engineer and architect, was a narcotic smuggler and weapons dealer protected by the Chief of Police in Tientsin's Italian Concession.(68) A suspected espionage agent used a dressmaking firm in Shanghai's French Concession as cover for both drug dealing and spying. A Soviet GPU agent worked in Harbin as a drug smuggler.(69)

The diversity of outside interests also embraced organized vice activities such as prostitution and gambling, and the classic underworld occupation of fence. The manager of Shanghai's "Anchor Hotel" and the owner of Harbin's "Nice Hotel" were both in the "white slave traffic."(70) A Greek drug dealer with a Chilean passport, resident of Shanghai's Savoy Apartments, owner of an import/export firm, was kept even busier organizing gambling houses for Chinese clients.(71)   Several dealers fenced an assortment of stolen goods.

Of all the collateral occupations engaged in by drug dealers, none was nearly so pervasive as the tobacco and fur trades. Neither item was per se illegal, but both were regulated and thus smuggled to beat taxes and quotas. A fair number of Russian drug smugglers covered or complemented their activities with an interest in furs. In Tientsin, D. Froiman imported furs, skins and leather products; H. Barkovich exported furs and carpets to the U.S.; D. Habinsky was a partner in the China Korean Fur Co. Ltd. with a branch in Harbin; and G. E. Kapustin thought to be the most notorious smuggler in Tientsin in 1933, owned a fur import and export business with offices in Dairen and Shanghai. In like manner, Greek dealers worked the tobacco trade in firms such as the United Tobacco Store, Express Cigarette Co., Tabaqueria Sino-Egyptiana, and the Tientsin Toabacco Co.

European criminal entrepreneurs active in China's underworlds during this period acted in typical ways. They weren't unique, although China surely was. China's political disarray and corruption, its boundless violence and drug use, all tied inextricably to European colonialism, Japanese imperialism, Russian (later Soviet) expansionism, and finally China's own split vision of modernity fated it to be a center for organized criminal activity on an unparalleled scale.   This extraordinary combination of circumstances presented organized criminals superb opportunities for gain, and they naturally rushed in.

World War I rendered several contradictory legacies. One was the attempt to rationalize the interactions of national governments on some reasonable bases, an attempt which failed when it was clear the League of Nations had no real, lasting power to discipline unreasonable nations. Controlling narcotic use through international agreements was League policy from the start and also failed for several reasons: the Chinese situation; European colonial systems (British reform nothithstanding) which paid many of their bills with revenues derived from regional drug trading(72); and the ability of European professional criminals to replace legitimate manufacturers of refined drugs. Regulation was also impeded because little could be done to limit access to opium (or coca) by criminal manufacturers. It was foolish for anyone to believe that nations which grew opium would really be amenable to limiting supply with so much demand. Opium was grown across a huge portion of the world's land mass, in places characterized by extreme rural poverty, tribalism, and antidelluvian land-tenure systems. Breaking through the local and transnational class systems which maintained opium agriculture was far beyond reformers' powers.

This era marked the time when illicit production and trafficking by Europeans took on more familiar contemporary dimensions. European professional criminals formed organizations whose nature was dynamic and fluid because the racketeers were ever alert to change and opportunity never hamstrung by custom. The criminals who took part in the transformation of the international traffic in drugs by migrating to the manufacturing end represented a variant of modernity itself. They were relativistic, lurid and urbane.

* * * * * * * * * * * * * * * * * * * *

NOTES

1. Among the finest accounts are Alfred McCoy with Cathleen B. Read and Leonard P. Adams, II, The Politics of Heroin in Southeast Asia (Harper & Row, 1972), and Alfred McCoy, Drug Traffic: Narcotics and Organized Crime in Australia (Harper & Row, 1980). The former is most enlightening when discussing narcotics in "post" World War II Indochina in particular C.I.A. involvement, while the latter is an absorbing examination of the Australian crime scene. The excellence of McCoy's work aside (and several other helpful studies cited below), there is such a general dearth of historical scholarship on drug underworlds and the activities of traffickers that researchers often are obliged to rely upon general surveys which contain information of some relevance to the issues of drugs and drug traffickers (see the comments on Romania in the text as an example) and journalistic accounts. The latter sources are particularly important when dealing with Chinese drug consumption and trafficking during the twenties and thirties.

Significant works which proved useful in understanding various segments of the drug phenomena included Arnold H. Taylor, American Diplomacy and the Narcotics Traffic, 1900-1939 (Duke University Press, 1969); Terry M, Parssinen, Secret Passions, Secret Remedies: Narcotic Drugs in British Society, 1820-1930 (Institute for the Study of Human Issues, 1983); David Courtwright, Dark Paradise: Opiate Addiction in America Before 1940 (Harvard University Press, 1982); Margaret Goldsmith, The Trail of Opium: The Eleventh Plague (Robert Hale Limited, 1939); and, J. M. Scott, The White Poppy: A History of Opium (Funk & Wagnalls, 1969).

2. There is no systematic analysis of Europe's drug underworlds, although there is some interesting impressionistic material about German and French organized crime. See H. Liang, The Berlin Police Force in the Weimar Republic (University of California Press, 1970), 145-147; and Otto Freidrich's marvellous Before the Deluge: A Portrait of Berlin in the 1920s (Harper & Row, 1972), 338-345. According to Friedrich, cocaine was Berlin's most popular drug in the 1920s. "It was everywhere," he commented, pushed by

everyone from "one-legged war veterans," to prostitutes, bar girls, artists, the artistically inclined, and doctors (p. 342). For those without ready access to underworld sources of supply, doctors were the most likely peddlers. Friedrich added that morphine was almost as fashionable and prevalent as cocaine.

Drug taking was part of the new Berlin "Zeitgeist," the vast public displays of sexual unorthodoxies seen at transvestite balls and pornographic shows. Experimentation with sex and drugs represented to some the collapse of all values. See Peter Gay Weimar Culture: The Outsider as Insider (Harper & Row, 1968), 129. Others, just as surely, found this sort of outrageous public behavior a naieve attempt at sexual and social experimentation which typically ended in shabby disappointment and a return to the norm. So playwright Erich Kastner wrote, "Some here, from sheer wish to be perverted/Found they had to the norm reverted"; Kastner, "Ragout fin de siecle," quoted in Susanne Everett, Lost Berlin (St. Martin's Press).

Germany was not the only nation in which crime, Bohemia, and narcotics merged at this time. France was another and some insight into this conjunction can be gleaned from the 1931 memoir of Alfred Morain, Prefet de Police, Paris; Morain, The Underworld of Paris: Secrets of the Surete (E.P. Dutton, 1931).

3. John T. Cusack, "Response of the Government of France to the International Heroin Problem," in Simmons and Said (eds.), Drugs, Politics, and Diplomacy: The International Connection (Sage, 1974), 230-31.

4. League of Nations, Advisory Committee on Traffic in Opium and Other Dangerous Drugs, Tenth Session, "Reply from the Government of Estonia to Circular Letter 8.1928 despatched on January 22nd, 1928," 208.

5. League, Advisory Committee, Minutes of the Fourteenth Session," Vol. I, Geneva, 9 January to 7 February 1931, 67.

6. League, Advisory Committee, "Analysis of the International Trade in Morphine, Dacetylmorphine and Cocaine for the Years 1925-1929, Appendix 6, LIST OF FACTORIES AUTHORIZED TO MANUFACTURE DRUGS COVERED BY THE GENEVA OPIUM CONVENTION (CHAPTER III), 403.

7. Twenty-one nations reported no primary manufacturing although all had pharmacies and many secondary firms concocting narcotic items for retail distribution. Nineteen nations reported drug factories but several pointed out they were only pharmacies. Poland and Austria both listed 23 factories but noted none of them produced opium, cocaine, morphine or heroin. Instead, they made tinctures and extracts. There was no discernible difference, therefore, between Roumania which claimed no manufacturing but identified

eleven laboratories preparing phials of cocaine or morphine and nations on the manufacturing list such as Austria, Poland, and Finland. Ibid., 404-411

8. Cusack, 232-33.

9. See League, op. cit., "Table III, Exports, Imports and Discrepancies, 1925-1929," 55-56.

10. League, op. cit., "Notes of Discrepancies in Table III (a)--Morphine," 86.

11. Ibid., 86.

12. Ibid., 88.

13. Ibid., 85.

14. Ibid., 90. The Swiss and Estonians had a squabble over 600 kilos of morphine which shows how complex, time-consuming, and often unsettling the regulatory process was. The Swiss claimed their exports were consigned to the "Progress Exportas-Importas Akcine Bendrave Kaunus" of Reval, and authorized under two permits (Nos. 26 and 95) issued by the Swiss Federal Public Health Service. The Swiss certificates followed receipt of two import authorizations given by the Estonian Consul-General in Berlin. Estonia replied that the Berlin consulate had no authority to authorize drug imports, and that it denied having done so. The competent agency was the Public Health and Social Welfare Administration of the Ministry of Education and Social Affairs located at Talinn. Also, the name of the alleged importing firm was clearly Lithuanian and translated into the Progess Export and Import Joint-Stock Company of Kaunus (Lithuania). The company neither existed in Kaunus nor in Talinn (also known as Reval). The Estonians added that Kaunus was a long way from Talinn and implied the Swiss knew the documentation was phony masking an illicit deal with drug traffickers. The Swiss stated the 600 kilos probably were "re-exported to Russia." Ibid., 92-93.

15. League of Nations, Opium Traffic Section of the Secretariat, "Convention for Limiting the Manufacture and Regulating the Distribution of Narcotic Drugs of July 13th, 1931: Historical and Technical Study," Geneva, October 1937, xx.

16. League of Nations, "REPORT TO THE COUNCIL ON THE PLAN DRAWN UP BY THE ADVISORY COMMITTEE ON TRAFFIC IN OPIUM AND OTHER DANGEROUS DRUGS WITH A VIEW TO THE LIMITATION OF THE MANUFACTURE OF NARCOTIC DRUGS," Geneva, 22 January 1931, 4.

17. Ibid., 5.

18. Ibid., 6.

19. Ibid., 7.

20. Ibid., 8.

21. League, op. cit., "Minutes of the Fourteenth Session," Vol. I, Geneva, 9 January to 7 February 1931, 67.

22. For the figures see Ibid., 13.

23. Cusack, 234.

24. League, op. cit., "Minutes of the Fourteenth Session," Twenty-Fifth Meeting (Public), 27 January 1931, 113.

25. An interesting example of this involved a triangular trade between European manufacturers, Central American smugglers, and U.S. drug importers and dealers. This was the Honduran connection which began sometime in 1932. Large shipments of almost pure heroin were purchased by a Honduran pharmacy from manufacturers in France, Switzerland, and Germany whose portion came from E. Merck in Darmstadt. The total amount was calculated to be more than enough for 40 years of Honduran medical needs. Destined for sale in the Gulf port of New Orleans, the heroin was owned by the former auditor of the Honduran National Railways. Eventually, four Honduran drug rings (also busy in counterfeiting and smuggling aliens to the U.S.) joined forces to sell the heroin. Part of their profit was used to purchase weapons needed to overthrow the Honduran government. The gangs' reported intent was to take over Honduras to facilitate their flourishing criminal activities. The coup d'etat was thwarted by U.S. drug agents who arrested most of the plotters. U.S. Treasury Department, Bureau of Narcotics, Traffic in Opium and Other Dangerous Drugs (Government Printing Office, 1936), 24-26.

26. This trade was even better known by the general public after the 1935 publication of a classic work of travel literature by adventurer and part-time smuggler Henry de Monfreid entitled Hashish: True Adventures of a Red Sea Smulggler in the Twenties. The work was reissued by Penguin Books in 1985.

27. See U.S. Treasury, Federal Bureau of Narcotics, Annual Reports, during the 1930s.

28. Eugen Weber, "Romania," in Hans Rogger and Eugen Weber (eds.), The European Right: A Historical Profile (University of California Press, 1966), 501.

29. Weber, 502.

30. Weber, 502.

31. Weber, 511.

32. Weber, 529-30.

33. Weber, 529.

34. Weber, 568.

35. Huri Islamoglu and Caglar Keyder, "Agenda for Ottoman History, REVIEW (Summer 1977) Vol. I, No. I, 41.

36. Islamoglu and Keyder, 41, n. 28.

37. Even the vertical integration of the illicit traffic didn't produce stable elaborately structured European organizations.

38. Egyptian Government, Central Narcotics Intelligence Bureau, Annual Report for the Year 1929 (Cairo: Government Press, 1930), 1.

39. Egyptian Government, 3-8.

40. Egyptian Government, 4.

41. Egyptian Government, 16.

42. Egyptian Government, 20.

43. Egyptian Government, 11.

44. Egyptian Government, 11-12.

45. Egyptian Government, 12-13.

46. The popularity of European morphine and heroin with Chinese consumers was the leading reason for Japan's entrance on the drug scene. In preparation for this, Japan sent students to Europe to study pharmaceutical chemistry in Europe. By 1927 (as noted above), they established a heroin factory in Istanbul which smuggled to Japanese distributors in Asia. In time, Japanese heroin was manufactured in Osaka, Japan, and forwarded to Chinese ports for distribution to the swelling mainland market. Soon after, Japan opened factories on Chinese soil (first in Dairen) as part of its military push onto mainland China. For their new Asian factories the Japanese needed "base-morphine" which they then converted to heroin. French factories initially supplied it. In the early 1930s, Istanbul manufacturers took over. Egyptian Government (1932), 21.

Soon, though, the Japanese made their own base-morphine allowing them independence from European manufacturers. By the time that happened, however, a new manufacturing competitor appeared. This was China which sought to vertically integrate all parts of the narcotic process. Chinese and Japanese drug competition formed an important part of the titanic struggle for the physical control of China which so marked Asian affairs for decades.

47. Papers of Harry Anslinger, U.S. Commissioner of the Federal Bureau of Narcotics, in Historical Collections and Labor Archives, Pattee Library, Pennsylvania State University Libraries, Box 10, File 9, "Result of investigations made in Athens by Bimbashi T. Marc," Inspector, Cairo City Police, 9/9/32, 8.

48. Egyptian Government (1933), 5.

49. League, op. cit., "Minutes of the Fourteenth Session," Thirty-Fifth Meeting (Public), 3 February 1931, 172.

50. League, 12.

51. Egyptian Government (1933), 16.

52. Staff and Editors of NEWSDAY, The Heroin Trail (New York: New American Library, 1974), 74.

53. Papers of Harry Anslinger, Box 8, File 9; undated and unsigned note.

54. Egyptian Government (1933), 18.

55. On trafficking in China see Johnathan Marshall, "Opium and the Politics of Gangsterism in Nationalist China, 1927-1945," Bulletin of Concerned Asian Scholars (July-September 1977) 8:3.

56. China's contribution to the world's traffic in morphine and heroin, run mainly through the foreign concessions, has quite likely been underestimated. Sterling Seagrave in his landmark The Soong Dynasty (Harper & Row, 1985 [Perennial Library Edition, 1986]) claimed an overwhelming percentage of the world's heroin came from China by the latter 1930s, much of it smuggled into America through Chinese diplomatic channels. This, however, seems to contradict an older argument that Japan had become the world's premier heroin supplier during that increasingly bellicose decade. Estimates of Japanese production and dominance must be re-thought or Seagrave's notions about China's preeminence downplayed. It is all a bit odd especially when it appears incontrovertible that Japanese factories inundated Chinese areas under their domination with tons of morphine and heroin. They exported to other world markets as well. Japanese drug factories on China's mainland used opium grown in north China, a region they successfully overran in the early 1930s. This placed them in direct control of Jehol province, a vast

opium-producing territory. That victory helped ensure, Seagrave himself wrote, Japan's "very profitable international heroin trade." Seagrave, 334-335.

57. Seagrave, 331.

58. Seagrave, 330.

The general desolation wrought by opium in China was painfully clear. In the interior provinces in the 1930s, it was pitiless. When the Communists went on their Long March in 1934 to escape annihilation by Chiang's forces, they shifted first to the west marching deep into the interior, to "frontier" provinces such as Guizhou. "This was opium country," states historian and journalist Harrison E. Salisbury in The Long March: The Untold Story (Harper & Row, 1985), where almost everyone from the age of fifteen on smoked opium--"They sat outside their huts puffing their pipes with glazed eyes, men, women, and teenagers" (p. 106). Guizhou's naked peasants owned nothing, were in debt from birth to some of the world's poorest landlords, sold their children to anyone willing and able to buy them, routinely practiced female infanticide and now and then in complete desperation killed male babies. Guizhou was, however, "saturated" with opium which "piled up in brown stacks in the sheds like cow dung put to dry" (Ibid). Opium was Guizhow's money and misfortune.

59. I am indebted to historian Terry Parsinnen for the information on Shansi.

60. Seagrave, 334.

61. Seagrave, 335.

62. Seagrave, 336-38.

Tu's power was, it seemed, limitless. He was a sort of Shanghai city councilor, chairman of the Chinese rate-payer's association in the French Concession, head of the Chung-Wai Bank and several others, directory of charity organizations, schools, businesses and labor unions. Among Tu's specialties was arbitration between labor and capital, and his decisions in these matters is considered completely binding. See Papers of Harry Anslinger, Box 10, File 3, "Tu Yueh Sheng and Opium Traffic," Annex - III, 3.

63. This particular rail line was one key to Japan's dominance in North China which by 1932 extended to almost all of China north of the Great Wall. During the course of the 1920s, Japan infiltrated vast stretches of Manchuria (both Japan and Russia wanted control of this huge region, around 413,000 square miles, which had very rich and large deposits of coal and iron. A particular method Japan employed in its plan was the stationing of troops on Chinese soil in order ostensibly to protect industrial projects like port

facilities and railroads. The Chinese previously conceded that chronic instability led them to accept such arrangements. Japan placed its "Kwantung Army" along the South Manchurian Railway claiming it was necessary to protect the railway from bandits. The Army then staged an incident, asserted Chineses bandits were responsible, and occupied Mukden. Japan subsequently took control of all the major south Manchurian cities.

64. Papers of Harry Anslinger, U.S. Treasury Department, American Consulate General, Shanghai, China, Office of the Treasury Attache, "Survey of Narcotic Situation in China and the Far East," 12 July 1934, Box 10, File 3, especially Annexes XIX-XXII.

65. Office of the Treasury Attache, Annex XIX, 1.

66. Office of the Treasury Attache, Annex XIX, 2.

67. Office of the Treasury Attache, Annex XXI, 3.

68. Office of the Treasury Attache, Annex XX, 8.

69. Office of the Treasury Attache, Annex XIX, 1, and Annex XXII, 3.

70. Office of the Treasury Attache, Annex XXII, 2, and Annex XIX, 2.

71. Office of the Treasury Attache, Annex XIX, 2.

72. See McCoy, and Seagrave.

PART III:
NEW VENTURES

Chapter 5
Organized Crime, Garbage, and Toxic Wastes: An Overview

**Published in U.S. Senate, Permanent Subcommittee on Investigations,
Hearings: Profile of Organized Crime, Mid-Atlantic Region (1983)**

Mr. Chairman and members of the Permanent Subcommittee on Investigations, my name is Alan Block and I am an associate professor of Criminal Justice at the University of Delaware. For approximately the past ten years, I have been conducting research and writing on the history, sociology, and criminology of organized crime. My particular concentration has been on the New York metropolitan region.

In 1981 I was doing research on waterfront organized crime when it became apparent that a number of the professional criminals and organized crime syndicates active in the Port were also deeply involved in the toxic waste industry. I subsequently focused my research on this new, to me, area of organized crime involvement. What follows is a summary of my research findings. As will be apparent, there are still numerous questions to pursue of which probably the most significant concerns the actual extent of organized crime's domination of the toxic waste industry and the accompanying political and criminal justice corruption. I might add, that I presume there is no point in my attempting to establish the general significance of toxic waste disposal as a social, indeed political, issue of enormous magnitude. The health hazards are well known. There are already so many reports and statements concerning environmental degradation and community health deterioration, that anything I could add would be redundant.

The areas I will cover fall into four general categories: (1) the connections between the solid waste disposal industry and the toxic waste disposal industry; (2) the presence of organized crime in first solid waste and then toxic waste; (3) the appalling overall futility of enforcement efforts concerning organized crime and toxic waste because of a malign combination of ineptitude and corruption; and (4) a few remarks on the potential national dimensions of the problem.

THE WASTE INDUSTRY

In many ways the toxic waste industry is the product of law creation. Prior to the 1970s there was no toxic waste industry as such--there was simply the waste industry within which certain firms may have specialized in hauling and disposing of toxic wastes to the exclusion of other forms of waste. But that type of specialization was not the result of particular technologies, nor the recognition that toxic wastes were inherently dangerous and demanded special treatment. Instead, it resulted simply and unintentionally from market pressures and opportunities to enhance profits. Even though the generators of toxic waste had been aware for quite some time that certain wastes were highly hazardous, they handled these substances, more likely than not, in traditional ways. It took political, indeed ultimately congressional action to define toxic waste disposal as a special field. Only with the general recognition that some waste was highly toxic and should, therefore, be subject to special rules and regulations did a new industry appear. The point is that the toxic waste industry was created primarily by State action, and that evidence indicates that it was and remains a sub-field of the waste industry in general. It appears that solid waste collection and disposal are and have been inextricably linked. It follows, then, that if organized crime is involved in the solid waste industry it is involved equally so in toxics.

Let me give an example, first, of the claim that solid waste firms and toxic waste firms are intimately related. One of the easiest and quickest ways to establish the connection is to glance at SCA Services, Inc., which is one of the largest waste service operations in the nation, and has already received a great deal of attention in the House of Representatives. According to Standard and Poor's Corporation Report of July 13, 1981, SCA operates in 32 states and the District of Columbia, and its activities include the collection, storage, transfer, processing, and recovery or disposal of solid or chemical wastes. In its first years of operation in the early 1970s, SCA acquired approximately 130 solid waste companies. The growth of SCA as well as the other major hazardous waste firms primarily came about by purchasing what were in effect garbage businesses. The following was a typical SCA announcement of this practice (located in the files of the New York State Senate Select Committee on Crime):

> BOSTON, Massachusetts, March 23, 1972--SCA Services, a Boston-based company specializing in solid waste management, building maintenance, and security/guard services has acquired two New Jersey solid waste management companies, Industrial Haulage Corp. and Avon Landfill Corp., it was announced today by Berton Steir, SCA President.

The New Jersey-based companies will continue to be managed by the former chief executive officers and owners, Louis Viola and his sons, Thomas C. Viola and Frank F. Viola. In addition to the acquired companies, certain assets of Intercity Service, Inc. (also a Viola-owned company are also being acquired by SCA.

The acquisitions were approved by the Public Utilities Commission of the State of New Jersey. The terms were not disclosed.

It was noted that the Viola family has been engaged in the solid waste disposal business in the State of New Jersey since 1911. In this connection, Intercity Service, Inc., provides municipal solid waste management while Industrial provides solid waste disposal for industry. Avon Landfill offers disposal management services.

The three companies combined were noted to have annual sales in excess of $2 million and operate 25 trucks servicing a population in excess of 150,000 in various communities in Northern New Jersey. Intercity Service provides municipal waste collection service to over 70,000 homes. The Viola companies are also considered leaders in private containerized solid waste removal services for Northeastern New Jersey.

According to Congressman Gore in the 1980 House Hearings, the crucial year (at least in New Jersey) when firms had to distinguish between garbage and toxic waste was 1978 when certain state requirements concerning hazardous waste disposal (basically a state licensing system accompanied by a manifest for tracking hazardous waste from generation to final disposal) went into effect. At that time, companies that had been generating both non- and hazardous waste "all of a sudden had to differentiate between garbage and toxic waste." And it followed, Gore stated, that the disposal companies that "had been disposing of their garbage now had to dispose of their toxic waste as well."(1) In the same House Hearing, one of the witnesses was Harold Kaufman who had unique experience in the toxic waste industry. Kaufman testified on a similar point stating that toxic waste haulers had been and many continued to be traditional garbage haulers. Speaking specifically about New Jersey and the installation of the State's manifest system Kaufman said:

> ... when they set up the manifest system in New
> Jersey, the way you got a hazardous waste license
> to haul was the simplest way in the world. You
> sent the State of New Jersey $50, they sent you
> a sticker, for everyone that you had on the PUC
> [Public Utilities Commission]. They made every
> garbageman in New Jersey that wanted to be, or
> is, because I got a lot of them their licenses, they
> made them hazardous waste haulers. Nobody
> checked out if they were qualified, did they know
> what they were doing, did they know what toxic
> waste was.(2)

Without doubt, there is an inextricable link between the garbage
industry in general and toxic waste firms in particular. State action
recognizing that toxic waste must be handled in a different manner than
traditional garbage did not rectify the environmental situation as planned, but
instead compounded the problem by making it more profitable for waste
firms to handle toxics. At the same time as state action drove the price of
handling toxic waste even higher, small and middle size generators
increasingly turned to those firms called "midnight dumpers" which undercut
the soaring prices because they merely hauled the toxic waste from the
generators and dumped it in streams, fields, sewers, landfills, etc. By
defining toxic wastes as a special item with escalating costs, State action
made this segment of the waste industry endlessly attractive for garbage firms
which had absolutely no facilities for handling toxics. It forced generators to
deal with highly unscrupulous haulers. Ultimately, though, the reasons why
State action in the toxic waste area have been so ineffective, are the
ineptitude and corruption in the area of regulation and enforcement. The
State created a framework for regulating toxic waste, but, so far, has shown
itself both unwilling and incapable of implementing it.

ORGANIZED CRIME AND GARBAGE--NEW YORK

Clearly, toxic waste disposal is a sub-field of the waste industry, and
the waste industry itself appears over time to be one of the sub-fields of
organized crime. Beginning in the mid-1950s, and continuing on since then,
evidence detailing the roles of organized crime in waste disposal have been
periodically presented. One of the most comprehensive and important
presentations of this relationship came in the fall of 1957 when the Senate
Select Committee on Improper Activities in the Labor or Management Field,
Chaired by Senator John McClellan, held hearings on the private sanitation
industry in New York. At that time, the industry did approximately $50
million worth of business annually serving 122,000 individual businessmen
and around 500,000 private homeowners. The major point of those hearings
was to show that organized crime was constructing "business empires in the

private carting industry through a system of monopoly enforced by trade associations and cooperative labor unions."(3)   In the main, the hearings focused on the activities of organized crime figures and associates, including Vincent J. Squillante, Bernard Adelstein, Joseph Parisi (who though dead in 1956 was important in the construction of the organized crime conspiracy under investigation), and Nick Ratteni.

Most of the testimony concerned Teamster Locals 27 and 813 which were under the domination of organized crime and were instrumental in furthering extortion in and from the private sanitation industry.   The Committee concentrated on Yonkers, New York, and Nassau County, Long Island.   In Yonkers, the focus was on Westchester Carting Co., which had been directed by Nick Ratteni since 1949.   Ratteni had an arrest record stretching back to 1926 and had served a considerable amount of time in Sing Sing Prison.   Among Ratteni's associates, it was pointed out, were the notorious Frank Costello, Anthony "Little Augie" Carfano, John Dioguardia, John Biello, and Joseph "Stretch" Stracci.

The Committee shifted attention from Westchester County to Long Island and narrowed in on he notorious Vincent J. Squillante who was the leader of the Greater New York Cartmen's Association and Bernard Adelstein the head of Teamster local 813. The Committee documented the ways in which both Squillante and Adelstein extended their influence into Nassau County, Long Island.   Included in this takeover was Squillante's nephew Jerry Mancuso. Chairman McClellan summarized this portion of the hearings stating that "Not only does Vincent J. Squillante, a hoodlum labor relations man, play a part, but we find that garbage collection industry men banded together in associations which eventually, under Squillante, invoked monopoly and restraint of trade arrangements with a system of punishments for nonconforming members through the use of whip companies."(4) McClellan stated that Bernard Adelstein who was the Secretary-Treasurer and Business Agent of Local 813, Private Sanitation Workers, worked together with Squillante in the scheme.   Adelstein compelled private sanitation firms to join Squillante's trade association, while Squillante induced "the cartmen of the Intercounty Cartmen's Association of Nassau and the Suffolk County Cartmen's Association to join the union, thus increasing the income and power of Bernie Adelstein."(5)

In his statement concluding the hearings, McClellan pointed out that "more than 46 hoodlums" were found "connected in one way or another with the carting industry" scrutinized by the Committee. Returning to Squillante, he remarked that he was "the self-styled godson of gangland executioner Albert Anastasia" and that he had no previous experience in anything except the "policy rackets and as a pusher of narcotics." Reiterating the evidence presented, McClellan said "that Squillante traded on his associations with the underworld and the union to:

(a)   Establish himself as the executive director of three separate employer associations.

(b)   Force individuals into the various associations and into Local 813, IBT.

(c)   Create a monopoly with respect to the collection of garbage and refuse in the Greater New York area.

(d)   Uphold and enforce the principle of territorial rights.

(e)   Trick the members of these associations into paying his back income taxes."(6)

Among the firms owned by organized crime figures identified in this investigation were the following:

> Sanitary Haulage Corp. (Anthony Ricci);
> Sunrise Sanitation (Carmine Tramunti and Anthony "Tony Ducks" Corallo);
> General Sanitation Co. (Nunzio Squillante and Vincent Squillante);
> Corsair Carting Co. (Nunzio Squillante and Vincent Squillante);
> Westchester Carting Co. (Nicholas Ratteni);
> Carters Landfill, Inc. (Vincent Squillante, James Licari, and Joseph "Joey Surprise" Feola);
> Jamaica Sanitation Co., Inc. (Gennaro "Jerry" Mancuso, Alfred "Pogy" Toriello, Joseph "Joey Surprise" Feola, Anthony "Little Augie" Carfano, and Frank Caruso).

The continuing involvement of the Squillante family in organized crime and the waste industry was also noted in an Intelligence Report compiled in 1968. At that time the focus was on Nunzio Squillante the brother of Vincent. The report states:

> Nunzio Squillante is the brother of Vincent J. Squillante, alias "Jimmy," who is wanted for bail jumping since March 7, 1961, on charges of Extortion and Attempted Extortion, date of arrest November 22, 1957. Vincent Squillante was last seen on September 30, 1960. Nunzio Squillante was also arrested on November 22, 1957, on an indictment charging Extortion and Attempted Extortion . . . . Testimony before the U.S. Senate Investigations Subcommittee on September 25,

> 1963, indicates that he was kidnapped and
> murdered on September 30, 1960. . . . Subject
> was formerly a principal with . . . Anthony
> D'Agostino in D and S Service Company. In
> 1967, D'Agostino and subject were reported
> organizing a rubbish removal firm called Flannary
> Towing Corporation, using a tugboat and three
> steel barges to burn refuse at sea. . . .
> Information indicates subject and D'Agostino are
> now operating New England Carting Company,
> Incorporated, . . . Hartford, Connecticut.

Just as the Squillante family (minus Vincent who had been murdered, of course) continued to influence the waste industry, so too did Ratteni who turned up once again in 1972 when the Newburgh <u>Evening News</u> reported that he was one of the principal organized crime figures attempting to extend racketeer influence in Rockland County over the private sanitation industry. According to reporter John McKeon, the complex situation in 1972 contained the following elements:

> Organized crime, in its effort to carve out a base
> of operations in the mid-Hudson area, has
> infiltrated the sanitation business, . . .

> Teamster Official THEODORE G.
> DALEY recently outlined a series of threats on
> his life which he feels are connected to a
> movement to take over sanitation operations on
> Teamster-manned construction sites.

> There is $1.5 billion worth of construction
> going on in this area, an official of LOCAL 445
> said recently. Access to these sites can be
> lucrative for an illegal operator.

> Several sanitation firms are under
> investigation for suspected links with organized
> crime.

Basically, the situation in the Newburgh area concerned an attempt on the part of Teamster Local 531 under the direction of Carmine Valenti and with the backing of Samuel Provenzano (brother of Anthony "Tony Pro" Provenzano), and Ratteni to replace Teamster Local 445 which was headed by Daley. Apparently, Daley had led an insurgent movement in the Teamsters in the mid-1950s which successfully ousted an earlier racketeer regime.

In enumerating the sum of Squillante's accomplishments, Senator McClellan described the criminological parameters for racketeering in waste. In almost all cases of industrial racketeering, one finds both corrupt trade associations and trade unions. Control of both employers and workers allows for innumerable methods of enrichment. Moreover, an industry organized in such a fashion is constantly engaged in restraint of trade, price fixing, and other forms of illegal activities. Add to that the fact that organized crime figures also owned firms in the waste industry, and imagine then the leverage available for those firms to dominate aspects of the market. In the waste industry not only were firms organized around trade associations and workers around trade unions, but as McClellan noted in section (d) of his summary, the entire industry is structured around "the principle of territorial rights." And one of the key tasks performed by organized crime was the enforcement of this principle. This means, naturally enough, that disputes over territories would be adjudicated by organized crime. At the very least, it is certain that the coercive elements in this industry were and remain substantial.

ORGANIZED CRIME AND GARBAGE--NEW JERSEY

Although the Senate Committee concentrated its energies on the waste industry in New York, a similar situation existed in New Jersey. In 1959, a New Jersey Senate Committee chaired by Senator Walter H. Jones held hearings on the waste industry in the Garden State. The Committee concentrated on the allocation of territories and how that principle manifested itself in bidding on municipal contracts. Three overlapping elements appeared in the testimony: constant allegations that bids were rigged; that municipal politicians were paid off; and that disputes were rife in the industry as firms struggled to bypass the territorial understandings and develop new lines of influence. On the seventh day of the Public Hearing (April 20, 1959) Vincent Ippolito (the Ippolito family had been in the garbage business in New Jersey for several decades and included several generations in partnerships) testified in the following way:

> Q.   Was there any mention of Fornaby?
> A.   Well, the Iommetti boys said they don't bid   Fornaby's jobs so Fornaby don't bid their jobs.
> Q.   Mr. Ippolito, during your conversation with the Iommetti's, isn't it true that there was a discussion about Fair Lawn and a payoff?
> A.   Yes, there was.
> Q.   And didn't Marangi say that in Fair Lawn there had   been   a   payoff   of   the municipal officials in connection   with the garbage bid? .. .

A.   Yes, he did say that.

Ippolito stated that payoffs to municipal officials were part of the customary arrangements in Hoboken as well. He acknowledged, in the same line of questioning, that Frank Stamato, a large waste entrepreneur in New Jersey (the Stamato clan, like the Ippolitos, Iommettis, Violas, and others, had a number of garbage enterprises), spent about $100,000 in payoffs to secure a municipal contract several years before in Hoboken. Apparently, Stamato factored in the payoffs in his successful bid which meant that corruption costs would be passed on. In conjunction with questions about payoffs, Committee Counsel also raised the issue of territories.

Q.   Did he give you any impression that payoffs were something that were a part of the activities of the   group, or the association?

A.   Well, he said that they had to give up jobs.

Q.   He said that the association had to give up jobs?

A.   Yes.

Q.   By that you mean the members of the association had to give up jobs?

A.   That's right.

Q.   And he said that members of the association had to   give up jobs because the people who were running the association were assigning the jobs to certain association members. Isn't that right?

A.   Yes, it is.

Q.   So that he told you that there was a complete system associated with bidding, which system selected--which system involved the selection of the bidder to get the job by the association and complementary bids by the other members of the association.

A.   That's right.

Ippolito went on to state that Crescent Roselle, one of the major waste operators in northern New Jersey (and again part of what were family businesses), offered Ippolito money if he would not bid on a job in East Paterson, New Jersey. The job had been fixed for Iommetti and the association did not want Ippolito involved. Roselle told Ippolito that "for $500, if you didn't bid that Iommetti would arrange for a complementary bid higher than his own and that the job would go to Iommetti."

The next line of questioning revealed another part of the intrinsic problems in the waste industry. Waste firms not only need customers with waste to purchase their services, but they also need locations to dump the waste. Access to dump sites is an indispensable element; to control dump sites such as landfills is to control the industry. In Ippolito's testimony both his particular difficulties and the general proposition is clearly stated.

Q. Mr. Ippolito, did you attempt to get a dump permit from any other contractors prior to the Hoboken bid?

A. Yes.

Q. And to what contractors did you go?

A. We went to Danny Malanka--

Q. What did Danny Malanka say?

A. He would not give us a permit.

Q. Did you offer him the price that he asked?

A. No. When we were going to put the bid in Hoboken the first time, we went to Danny Malanka's office to get a permit to dump garbage at his dump and he refused us. He said he wouldn't give us the permit because we were from Bergen County and he said that he was going to run all the Bergen County fellows out of Hudson County.

Q. In other words, he would not give you a permit under any circumstances?

A. That's right.

Q. Did you speak to Stamato, prior to the Hoboken bid?

A. Yes.

Q. Why did you go to Stamato?

A. He said he would get us a dump permit from Viola.

Stamato, however, did not get Ippolito a permit because he himself was bidding on the job. The testimony of Vincent Ippolito was followed by that of his cousin Joseph who added that the Ippolitos were not allowed to join the Garbage Contractor's Association. Not being members meant, he stated, that the firm couldn't get dumping permits and, therefore, couldn't bid for jobs. control of the dumps was in the hands of the association, he testified, which included firms owned by Crescent Roselle, Thomas Viola, Frank Stamato, and Chester Iommetti, among others. Little more than ten years later, a number of the leaders of the association would be among the largest toxic waste disposers in the nation.

New Jersey, in the late 1950s, had, in addition to the association, an equivalent to New York's Local 813. This was Local 945 which witnesses testified, according to Bob Windrom's report in House Hearings, forced "members of the waste trade association . . . to contribute money yearly to buy the local's business manager, John Serratelli, a Cadillac."(7)   One result of the New Jersey hearings was Serratelli's indictment "on state charges" that garbage contractor, Alfred Lippman, "gave him $4,000 in bribes to secure labor peace." This indictment was swiftly followed by a second one charging that Serratelli "and others in the garbage industry arranged to rig bidding on the Belleville, NJ, garbage collection contract in 1955.(8) Unfortunately for Serratelli, he disappeared and was presumably murdered "for failing to go along with the directions of organized crime leaders who warned him against cooperating with authorities in any way."(9)   It seems that Serratelli like Vincent Squillante had become too great a liability for organized crime. Consider also that the primary upshot of the late 1950s investigations was the murders of Squillante and Serratelli--there were no structural changes within the waste industry, not even a pattern of prosecutions indicating that law enforcement was seriously monitoring the industry.

CONTINUING REVELATIONS

One indication that the waste industry as structured by organized crime was basically unaffected by past revelations is revealed by focusing again on Westchester County. In 1967, Charles Grutzner of The New York Times reported that "an investigation directed by United States Attorney Robert M. Morgenthau has found that 90 percent of the trade-waste disposal in Westchester County is handled by members of the Genovese and Gambino families of the Mafia" (6/28/87). Grutzner wrote that Morgenthau began his investigation following the presumed murder of Joseph "Joey Surprise" Feola who had "taken the waste-disposal contract of the Ford Motor Company's giant assembly plant at Mahway, NJ., from a Gambino-controlled waste removal company." Also identified as still active in Westchester waste was Nick Ratteni who, along with Sabato Milo, headed the Genovese crime syndicate operations in Westchester. About six months later, The New York Times added more detail to the story reporting the arrest of Sabato Milo's son, Thomas, "on charges of trying to intimidate a competitor."(10) In addition, it was pointed out that the private carting industry in Westchester County (as in New York City) had undergone a process of "tremendous consolidation" with "most of the lucrative business . . . going increasingly to four large companies that each preside over a well-defined section of the county." The four major companies identified were:

> (1)   Queen City Refuse Collectors, a Milo
>   operation;

(2)   Greeley Sanitation Service, Joseph and
        Robert Ligouri owner (no mention of
        organized crime connections);
(3)   Westchester Carting, operated by Ratteni;
        and,
(4)   Valley Carting Company, "run by Tobia
        De Micco and Nichola Melillo, whom law
        enforcement agencies say represent the
        interests of the Gambino family in
        Westchester."

A revealing portion of the story, reported by Ralph Blumenthal, concerned a conversation between Vincent M. Fiorello "operator of several carting concerns in upper Westchester, Dutchess County, the Bronx and New Jersey" and Tony Vone of New Rochelle Garbage Collectors. The talk consisted of Fiorello telling Vone that a trade of garbage stops that they had worked out involving the Geigy Chemical Corporation could not go through because of opposition from Ratteni and DeMicco.

The following decade was simply more of the same, except for the fact that the State recognized that some garbage was toxic and instituted procedures and rules to handle the disposal of this new category. A few years prior, however, another scandal emerged linking waste businesses to organized crime. This time the area scrutinized was Brooklyn. On March 28, 1974, Brooklyn District Attorney Eugene Gold revealed that "Fifty-five private carting companies in Brooklyn and nine carting industry officials" had been indicted on "charges of restraint of trade and perjury."(11) The District Attorney stated that "the carting industry in his borough was controlled by organized crime through the Brooklyn Trade Waste Association." Organized crime controlled what was a $60 million-a-year industry which included what Gold claimed were approximately $20 million-a-year in overcharges resulting from "collusive practices in the industry." In fact, the Brooklyn Trade Waste Association itself "had taken in fees of $100,000 or more for assigning private carters their territories in the borough."

The Brooklyn Trade Waste Association officials indicted were the President, Patsy D'Avanzo, Vice-President Sam Galasso, Joseph Schipani "who has been identified by law enforcement officials as a soldier in the Genovese organized-crime family," Joseph Dantuoro "who has been identified as a member of the Colombo organized crime family," Salvatore Sindone, Sabino Colluci, Dominic Colluci, Michael Russo, and John Cassillo. Gold added that Schipani had only been released from federal prison the year before after serving a sentence for income tax evasion.

THE MERGING OF TOXIC AND SOLID WASTE

Six years later on October 17, 1980, across the Hudson River in New Jersey "a statewide grand jury indicted two garbage industry trade associations, 24 corporations and 25 individuals, including some identified as members of organized crime, on charges of conspiring to control commercial garbage contracts in nine Northern New Jersey Counties."(12) The two trade associations were the New Jersey Trade Waste Association and the Hudson County Sanitation Association which between them represented "15 percent of the statewide garbage industry"(13) controlling "approximately 35 percent of the total garbage collection revenues in the state, or about $46 million annually."(14) It is significant that in the Star Ledger story by Peter Marks on the indictment, a number of firms and individuals involved in the toxic waste business are mentioned including ISA in New Jersey, Inc., of Mahwah and its president, Louis Mongelli; A. Rizzo Carting Co., Inc., of Wayne and its president, Anthony Rizzo; Statewide Environmental Contractors and its officers Charles Macaluso and Frank Lotano; and Duane Marine Salvage Corp. of Perth Amboy which had been indicted in September on charges of "dumping 500,000 gallons of toxic waste into the Perth Amboy sewer system."

But perhaps the most revealing part of the indictment for those interested in the issue of toxic waste and organized crime is in that section dealing with Browning-Ferris Industries of Elizabeth, New Jersey, Inc. First of all, Browning-Ferris was and may still be the second largest waste firm in the nation with a significant portion of their business being toxic waste. This past summer, Browning-Ferris offered to buy out SCA Services for $197 million, according to the Wall Street Journal. And while there is no mention in the indictment of the Elizabeth operation being a toxic waste one, it is nevertheless reasonable to suppose that the kind of activities carried out by the Elizabeth division were representative of the firm's methods overall. In the indictment, Browning-Ferris is described as being a principal "during the period of this conspiracy . . . in which the conspirators operated to restrain trade by enforcement of 'property rights.'"(15) The particulars in the indictment show Browning-Ferris as both a protector of the illegal property rights system, and initiating a "grievance hearing" settled in their favor by recognized organized crime figures in the industry.

From only the relatively few examples cited, there are some reasonable conclusions which can be drawn at this stage. From about the mid-1950s until today, the major trade associations in the waste business both in New Jersey and New York have been dominated by organized crime. Key officers and participants have been identified by many law enforcement agencies and legislative committees as organized crime figures. In addition, the methodology of business within the waste industry has been aptly characterized by the descriptive terminology of the many indictments and investigatory documents: it is an industry illegally controlled by organized crime in which business is illegally conducted. It is an industry in which organized crime not only establishes the parameters within which business is conducted, but one in which organized crime adjudicates disputes through its customary methods of intimidation and violence. The power of organized crime in this industry is immeasurably increased by its control over workers and non-organized crime firms through the domination of certain major Teamster locals. The ones most prominently mentioned are Local 813 and Local 945. By controlling the major trade associations, the primary trade unions, major waste firms, and landfills all with the help of friends in political and criminal justice positions, organized crime is effectively the waste industry, and it may not matter very much whether one is talking about solid, or toxic waste.

An example of the merging of organized crime, solid waste firms and toxic waste disposal was provided by David Colton of the Journal News, Nyack, New York, in the summer of 1980. Colton's report(16) comments that while charges of toxic waste dumping by organized crime figures in Orange County, New York, have surfaced, and been dismissed by Ralph Smith the "head of the State Organized Crime Task Force," an independent review indicates the strong probability of organized crime dumping toxics in the county since 1970. Among the sites mentioned were a "private landfill" in Warwick where the New York State Department of Environmental Conservation "has strong reason to believe that a lot of industrial waste was disposed of" by a waste firm named Grace Disposal which leased the property. Organized crime figure Louis Mongelli controlled Grace Disposal and another firm "which holds the contract to dispose of Ford's industrial waste." Eyewitness testimony claimed tractor-trailer trucks had unloaded industrial waste from the Ford Assembly Plant in Mahway, New Jersey. Another Mongelli company, All-County Environmental, was reportedly seen dumping at another site in the county. At that site, the State had licensed DV Waste Control to operate. It is especially interesting that DV Waste Control "is operated by Vincent DeVito," reportedly a convicted loanshark and the brother-in-law of Thomas Milo discussed earlier in a slightly different context. Milo and his firm, Suburban Carting, operated the Al Turi Landfill in Goshen, New York, and Colton writes, that landfill "is listed both by the state and in an October, 1979, federal Waste Disposal Site Survey as possibly containing hazardous chemicals."

The final part of Colton's story deals with the somewhat mysterious affairs of Jeffrey and Anthony Gaess, principals of Gaess Environmental Corp., and the C & D Disposal company, which was sold to DeVito's firm, the above-mentioned DV Waste Control. The connection to Gaess Environmental was made when toxic waste dumping around Harriman, New York, was discovered and two buried tanker trucks displaying the name of C & D Disposal were found. It was established that the trucks were actually registered to Gaess Environmental Corp., which was, in turn, a subsidiary of SCA Services. The last bit of bad news uncovered by Colton concerning Gaess was provided by the lessee of the Harriman property, John

Kalodziejski, who said "that about 20 barrels of industrial waste were removed from the site last spring," and then added that he had "sold his Harriman Sewage Disposal Company to Gaess Environmental."

More information on Jeffrey R. Gaess and especially his brother Anthony D. Gaess was located in an investigator's Report compiled for the New York State Senate Select Committee on Crime. Drawn from several different sources, it revealed that Gaess Environmental Services was purchased by the SCA/Earthline Division in 1977. Concerning the background of Gaess Environmental, it was learned that a firm called Tony Gaess Service Corp. was formed in the summer of 1966 in Bergen County, New Jersey, and six years later changed its name to Gaess Environmental. At that time, its officers were Anthony D. Gaess, President, and Christopher P. Recklitis, Vice President. Recklitis was later charged "by the Securities and Exchange Commission of diverting four million dollars of SCA's company money for his own use," with the aid of SCA President Berton Steir and SCA's northeast regional controller, Nicholas Liakis. Bob Dearing of the Buffalo <u>Courier</u> <u>Express</u> reported that the SEC investigation began in 1974, "and during the course of the three-year probe Recklitis" who had been SCA president from 1972 to 1974 "and Berton Steir, SCA founder and chairman, resigned from the concern." Working with Recklitis in the scheme was Anthony "Tony Bentro" Bentrovato "who had been indicted in 1975 in a kickback operation case involving the Upstate New York Teamsters Pension Fund, along with reputed organized crime kingpin Anthony 'Tony Pro' Provenzano."

Other information in the report indicated that Anthony Gaess was connected with Richard Miele "the proprietor of a firm known in the trade as Modern Transportation, although the correct legal entity actually is Sanitary Waste Carriers, Inc., of Kearney, N.J." Allegations of improper handling of, at least, municipal sludge, by Miele have surfaced in Philadelphia and Baltimore in the very recent past. In addition, a chart prepared by the New Jersey Department of Environmental Protection indicates that Richard Miele owns 25 percent of a business called MSLA which was a landfill in which toxic chemicals were dumped, according to police intelligence surveillance reports. Next to the MSLA landfill were C. Egan & Sons Sanitary Landfill and the P.M. Landfill. The procedure for illegally dumping at MSLA was simple, with trucks using gate #2 on the site which was reportedly opened by a guard who was paid for his service and "informed not to say anything." In addition to Miele, others with interests in MSLA included William and John Keegan, Joe Cassini, Mike Cignarella, and Crescent Roselle.

Earlier, Crescent Roselle was discussed in conjunction with the 1959 New Jersey State investigation of the waste industry. In 1973 and 1974, Roselle's various waste businesses, with one major exception, were bought by SCA Services. The particular Roselle firms bought by SCA were noted two years later when he wrote to Thomas Viola, by then vice president of SCA Services, stating that the following individuals were extending their options as stated in their employment contracts with SCA to continue in the employment of SCA "for at least the next ten years":

Peter Roselle, Jr.	Waste Disposal, Inc.
Joseph C. Roselle	Waste Disposal, Inc.
Peter Marinaro	R.R. & M.
Louis P. Roselle	Peter Roselle & Sons Co. & Fereday & Meyer Co., Inc.
Alfred J. Lippman	Roselle-Lippman Co.
Crescent J. Roselle	Peter Roselle & Sons Co.
Crescent J. Roselle	R.F. & M.
Arthur Roselle	Peter Roselle & Sons Co.

Crescent Roselle reminded Viola that "all of these original employment contracts are with the above companies, and all of these companies, as you know, are now known as Waste Disposal, Inc." According to police intelligence reports compiled by detectives in the Elizabeth, New Jersey, Police Department, Anthony Gaess met with Crescent Roselle on a regular basis in order to show Roselle how best to profit "in handling chemicals." Also, in the same intelligence reports it is claimed that Gaess "was caught stealing accounts and running drums of toxic waste into" the Kin Buc Landfill. On December 22, 1980, Crescent Roselle was murdered becoming the third waste entrepreneur to be murdered in just a few years.

Without intending to be melodramatic, let me identify some of the toxic waste generators who have done business with Gaess in the past (the list is drawn from Generator records on file with a Congressional subcommittee).

(1)	Reichhold Chemicals, Inc.	1971 - 1978
(2)	Borden Chemical, Division Borden, Inc.	1978
(3)	Air Products & Chemicals, Inc.	1975 - 1979
(4)	NL Industries, Inc., NL Chemicals Division	
(5)	Refined-Onyx Division Polyurethane Specialties Co., Kewanee Industries,	

	Inc.--Subsidiary of Gulf Oil	1970 - 1977
(6)	Diamond Shamrock Corp., Process Chem Division	1976
(7)	Diamond Shamrock Corp., Soda Products Division	1977
(8)	American Cyanamid Co., Lederle Laboratories	1970 - 1974
(9)	Millmaster Onyx-Divison of Gulf Oil Corp.	
(10)	DuPont Co., F & F Department	
(11)	Stauffer Chemical Co., Specialty Chemical Division	1976, 1978 - 1979
(12)	Stauffer Chemical Co., Plastics Division	1975 - 1978
(13)	Mobil Oil Corp., Mobil Chemical Company	1971 - 1973, 1977
(14)	Pfizer, Inc.	1974

One other reported instance needs mentioning. The American Cyanamid Company stated that from 1969 to 1973 they had contracted with the Bushbaum Sanitary Service to dispose of 200 tons of "process waste" which was taken to an unknown location. The company identifies Bushbaum Sanitary Service as an Anthony Gaess firm. American Cyanamid also stated that they had no idea what was contained in the waste.

I commented that the Gaess brothers were somewhat mysterious, or at least their activities were. Some of the known and some of the suspected Gaess activities have been enumerated already. However, it is worth pondering several others. Why is it, for instance, that New Jersey Department of Environmental Protection records dealing with Gaess are cross-referenced to R & R Sanitation of Randolph, New Jersey? DEP files indicate that R & R Sanitation is a "trade name for Carl Gulick, Inc. . . . a handler of industrial waste in Mt. Freedom, NJ, at one time." Intriguing as that may be for tracing some of the Gaess's hidden interests, it is more interesting that attorneys for SCA Services, SCA Services of Passaic, Inc., Earthline Company, and Anthony Gaess in a brief(17) denied that Gaess was the principal operating officer of Earthline. This despite the fact that minutes from a special meeting(18) of the Board of Directors of Scientific, Inc., held that "management had been investigating the feasibility of acquiring from SCA Service,s Inc., the Gaess group which consists of SCA's liquid collection business." It was reported "that the business was run by Tony Gaess aggressively with a view toward expansion;" that "Mr. Gaess would make an ideal coordinator of Scientific's landfill and commercial collection operations;" and it was understood that the previous year's "profits for the Gaess group were $750,000." The brief denying that Gaess was Earthline was filed in answer to a civil action #79-514 brought by the U.S.

Attorney for the District of New Jersey in which Gaess, SCA Services, and nine others were named for unsafe and improper methods in the handling and disposal of millions of gallons of liquid waste, much of it toxic, which was placed in the Kin Buc Landfill. If Anthony Gaess was not the operating officer of Earthline, he nevertheless did receive a $140,000 loan from SCA in the form of "an unsecured note which is not interest bearing. Payments are due at the rate of $14,000 per year, for 10 years, beginning on September 1, 1976. During fiscal 1978, SCA agreed to postpone the September 1977 payment to September 1978, in consideration of Mr. Gaess pledging 3000 shares of SCA common stock as collateral for the unpaid installment." Obviously enough, there are substantial unanswered questions concerning Gaess and almost all the other individuals and firms mentioned so far, including SCA, which must be addressed. And given the types of offenses committed or alleged to have been, answers to them all the more imperative.

THE POLITICS OF TOXIC WASTE ENFORCEMENT

Unfortunately, there is at the moment little hope that investigations can be mounted which will successfully face the peril and provide the answers so desperately needed. Certainly the record in the Northeast leaves so much to be desired that public confidence must be shaken if not shattered. Leaving aside for the moment the fact that organized crime has controlled the waste industry for almost thirty years at a minimum, despite repeated investigations and disclosures, and looking solely at the enforcement record dealing with hazardous waste violations regardless of whether or not the principals are organized crime figures or associates, one find an enforcement record which indicates either or both ineptitude and corruption on a monumental scale. In New Jersey, the most current example is the case of State Police Detective-Sergeant Dirk Ottens one of the most knowledgeable police officers on toxic waste and organized crime in the nation. However, following a subpoenaed appearance before a closed hearing of a House subcommittee investigating organized crime and the toxic waste industry and a subsequent public hearing, Ottens has found himself in a state of political/criminal justice limbo.

An editorial in New Jersey's Home News(19) noted that the recent Congressional report "which charges the state Department of Environmental Protection, the U.S. Environmental Protection Agency and members of the state Attorney General's Office with investigative bungling that allowed thousands of gallons of toxic waste to continue pouring into landfills years after the illegal dumps should have been closed," still nevertheless praised Ottens and his partner. Since the winter of 1980-1981, Ottens has been commanded to leave toxic waste alone and placed in some of the most demeaning and menial tasks the State Police could construct. The Home News asks whether these kinds of assignments "have any relationship to the testimony offered to a federal judge who concluded Ottens had developed hard evidence DEP officials were taking kickbacks from toxic waste firms?" If the answer is affirmative, it might explain why the identity of one of Ottens' most important informants in the toxic waste industry was revealed by a high official in Criminal Justice. This placed his life in jeopardy and made it that much more difficult for Ottens to develop leads. It would also explain and perhaps affirm another informant's claim that the State Police were trying to frame Ottens. It appears that it was Ottens' expertise and desire to follow the trail of organized crime and toxic waste into the arena of public corruption that was his downfall. Surely the fact that Ottens as well as other witnesses from New Jersey who appeared before the House subcommittee were told by state officials to stay away from the issue of public corruption raises the probability higher. The issue of Ottens' purgatory was serious enough to cause the chairman of the House subcommittee to ask the Attorney General of New Jersey to conduct a complete independent review of the matter, including an interview with Ottens. The request was made in the summer of 1982 and, so far as can be determined, never carried out.

Among the allegations that Ottens might have been able to pursue are claims that an official in the Department of Environmental Protection was simultaneously counsel to a toxic waste firm; that a former New Jersey Deputy Attorney General was an investor in Kit Enterprises, reportedly a firm involved in illegal toxic waste activity, and controlled by two major organized crime figures--JoJo Ferrarra and John Riggi; that a second former Deputy Attorney General owned part of Global landfill in Ocean County and that an investigation into this affair ended when another Deputy Attorney General received a sailboat; that the investigation into Duane Marine was marred by corrupt acts by officials in the DEP and New Jersey's Division of Criminal Justice; and finally, that certain officials in New Jersey's Public Utilities Commission received bribes by waste firms. Certainly, it is true that allegations of corruption in New Jersey concerning the waste industry are not new; rigged bids and payoffs to scores of public officials were a mainstay of testimony beginning in 1959 and continuing on ever after. It may well be, then, that the Ottens' affair is part of the traditional way in which political and criminal justice is carried out in New Jersey. What makes this more significant, however, is that this time it is not morality which is assailed, but the public health. It is one thing to take a cut from vice enterprises, and quite another to protect and profit from massive environmental poisoning.

New Jersey, of course, is not alone in subverting or bungling the criminal justice process when it comes to toxic wastes. It seems, for example, that certain governmental agencies in New York are withholding and suppressing vital information about toxic wastes from the public, while others are carefully editing material germaine to the issue of the manner in which toxics are handled and by whom, and much about how agencies are grossly mishandled environmental and health hazards.

For example, on September 7, 1982, the Staten Island <u>Advance</u> reported that "Arsenic, mercury, lead and other toxic metals seeping out of three city garbage dumps might be accumulating in shellfish and killfish growing in Jamaica Bay, a state panel of landfill experts wrote in a recent unpublished report." The <u>Advance</u> was referring to a study conducted by an ad hoc group known as the New York City Landfill Assessment Team composed of Earl H. Barcomb from the New York State Department of Environmental Conservation/Division of Solid Waste, Robert L. Collin from the Division of Waste, and Wallace E. Sonntag representing the Division of Air. In general, the ad hoc team found that all prior analyses of New York City's landfills conducted by the City Health Department and the City's Department of Sanitation were so flawed as to be worse than useless, and that the landfills did present a clear and present danger to the public health. In fact, the City's landfills have been used for years by toxic waste dumpers and the City agencies supposed to monitor the landfills were either so inept or corrupt that they couldn't or wouldn't act in the public interest. Also, the State Department of Environmental Conservation has yet to explain why the ad hoc team's report was not released, and coincident with this why the DEC fired Samuel J. Kearing who had been the Regional Counsel for Region Number 2 of the DEC, and had been trying for several years to get the DEC to "enforce the law against the City of New York." Kearing joins a list of individuals whose careers have been damaged because they seem to have posed a threat to powerful political and criminal interests which may well be combined in the toxic waste industry.

The official intolerance for so-called "whistle blowers" and aggressive investigators should be placed alongside the remarkable tolerance exhibited for the corrupt and criminal in the waste industry. Back in 1975, for instance, the Brooklyn Trade Waste Association, which had been ordered dissolved following the Gold investigation mentioned earlier, was, nevertheless, deeply involved in negotiations with the city over a strike in the private carting industry called by Teamster Local 813 which was still headed by Bernard Adelstein. Damon Stetson reporting for <u>The New York Times</u> noted that the city's Consumer Affairs Commissioner, Elinor C. Guggenheimer, and Leo Pollack, the head of Consumer Affairs Trade Waste Division, were both satisfied that there was "no evidence that strong-arm tactics were now being used in normal operations."(20) District Attorney Gold disagreed, according to Stetson, but the resurrection of the Brooklyn Trade Waste Association continued, despite the extraordinarily equivocal nature of the quoted statement. No one, it seems, asked what "normal operations" meant, or whether Consumer Affairs was actively investigating the industry or merely passively waiting for complaints to surface.

Given the violence associated with the industry, it does appear that waiting for complaints is probably the least effective method of monitoring. Nonetheless, waiting for complaints characterizes the agency's strategy.

Anyone in the industry in the Northeast during the last two decades or so would have been well aware of the murders of Vincent Squillante, John Seratelli, Joseph Feola, Alfred DiNardi, Gabriel San Felice, Crescent Roselle, and John Montesano who was one of the major witnesses for the Senate investigating committee back in 1957 and was killed in 1981. In fact, several years before his murder Montesano told Newsday reporters, Renner and Mulvaney, "You know there's a [murder] contract on me? You're the third one to tell me about it. Because I testified [in the Senate hearings], I was the heavy. I did the right thing . . . Now, I hear they're going to whack me out . . . leave my brains on the street."(21) Knowledge of these events is hardly likely to bring witnesses forward or to encourage others with information about industry illegalities to march down to Consumer Affairs.

Perhaps there were individuals who decided that reporting crime in the industry was the height of foolishness, because they were aware that the Drug Enforcement Administration "used a New Jersey toxic waste-disposal company with alleged ties to organized crime to dispose of discarded pharmaceuticals and other drugs in 1978."(22) Concerning this rather astonishing event, the DEA did state that the drugs taken to the Chemical Control Corporation facility were "all properly incinerated in the presence of drug-enforcement agents and could not have been diverted or reused," and that the agency was "unaware then of any criminal connections" to the firm. Questions have been raised, however, "by the arrest of one of the alleged company principals . . . on charges of financing a narcotics laboratory and a report that some undestroyed pharmaceuticals and drugs were discovered at the waste-strewn company warehouse before a suspicious fire" totally destroyed the facility in 1980 two years after the DEA used the firm. In addition, investigators have stated that "law enforcement files available to the agency before its use of Chemical Control had long contained information about alleged criminal associations of the company and John Albert."(23) One can only imagine the dismay potential witnesses or informants would experience, if it was common knowledge that the DEA utilized Chemical Control and John Albert for the alleged incineration of narcotics.

It seems fair to say that a summary of activities and events in the waste industry over the past quarter century, and the toxic waste industry since the late 1970s when it was created, displays an unmistakable pattern of ineptitude or corruption. Serious and capable investigators are harassed, intimidated, indeed dismissed; reports are erroneous or, if accurate, suppressed; evidence in cases is mishandled; criminal organizations are legitimized by municipal officials; the identity of informants is revealed; state grand jury proceedings have been used to prevent pertinent information from reaching Congressional investigators; and clear conflicts-of-interest between regulators and the regulated are permitted.

What makes all the above even more illuminating, if not menacing, however, is that New York and New Jersey have been among the most carefully scrutinized of states. Probably no area in the nation has received as much attention as the New York metropolitan region, and that includes attention from those interested in organized crime. Add to that the fact that New Jersey is supposedly a bellwether state in the field of toxic waste investigation and enforcement, and the scope of the national problem emerges. If, as a number of individuals claim, New Jersey at least does something about toxic waste through its manifest system and its Hazardous Waste Task Force, can one imagine the plight of the rest?

PROBLEMS IN THE SUNBELT--FLORIDA, NEW MEXICO

As early as the summer of 1973, the State of Florida somewhat tentatively began collecting information on the garbage and recycling industry because of the "possibility of organized crime infiltration and/or monopolistic tendencies." In particular, the Florida Department of Law Enforcement was interested in two SCA subsidiaries and canvassed other states for information which revealed the following:

> Tri-County Sanitation, Incorporated, Detroit, Michigan, had been purchased by SCA in September, 1972. Tri-County Sanitation had been monitored by Michigan authorities since 1962, when Joseph Barbara, son of the host of the Appalachian meeting of top organized crime leaders (1957), and Paul Vitale, brother of Pete Vitale, identified [organized crime] figure, formed the company. The Vitale group was still in control of Tri-County when purchased by SCA.

> The California Department of Justice revealed that SCA purchased G.I. Ecological Waste Association, Incorporated, in late 1972. Manuel Assadurian served as an officer in G.I. Ecological. Manuel Assadurian and his brother, Sam, have been associated with numerous California garbage companies as well as Louis Visco, an organized crime associate, and Luigi Gelfuso, another organized crime associate.

Included in the Florida report is a section on various Florida counties compiled because of "allegations that organized crime and New York companies" are taking over garbage collection. In Dade County firms investigated included Packed Sanitation Service, Inc., located in Hialeah, Florida, whose officers included Joseph Pelose, a reputed associate of Nick Ratteni and one of the original owners of Rex Carting Co. discussed earlier. Other associates of Pelose include the alleged organized crime figure John Masiello. In Palm Beach County the president and vice president of Waste Compactor's, Inc., were reportedly associates of Gerardo Catena a major organized crime figure from northern New Jersey living in Florida. Information on the two officers "reflects that . . . [they] left their former residence in New York during a grand jury investigation on bookmaking." Investigation in Pasco County produced an even more interesting organized crime connection. Both the West Pasco Disposal Company and Community Disposal Services, Inc., were "controlled by Joseph Messina and his brother Salvatore . . . [who] are associated with James Failla, a capo in the Gambino LCN 'family' and chauffeur to Carlo Gambino. . . . Recent information reveals that Failla and a Philip Modica visited the Mesina's in New Port Richey." In 1976, corroborating intelligence on these relationships was developed in Suffolk County, Long Island. Centering on an individual named Tommy DeSousa it was learned that he "is an enforcer for the Carlo Gambino crime family in the James Failla regime. His immediate boss through wiretap [information], is known as 'Phil,' later known as Phil Modica who is one of three brothers who are all soldiers in the Gambino family. . . . DeSousa in 1975 was in constant telephonic touch with Community Waste Disposal. . . . Conversations clearly disclose that DeSousa was giving orders" to the Florida operation "about fixing prices."

In somewhat the same manner that Florida experienced the expansive efforts of New York metropolitan region organized crime figures in the waste industry, so apparently has New Mexico. According to police intelligence files, several New York based organized crime figures have "put pressure of local carting businesses to buy them or scare them out of business, in the Albuquerque area." A summary of intelligence up to the mid-1970s on this situation found the sale of a local waste operation to Joseph Capone, Jr., and Gary Curreri both from New York. One of the intelligence reports located in the files of the Kings County District Attorney's Office then state:

> Source in New Mexico said that they were sent there by . . . an underling of Joseph Pagano, Genovese Family Capo who operates carting firms in Westchester and was involved . . . in a Medicaid Fraud Scheme. Pagano was also tightly linked in previous stories with the New York Purple Gang, a narcotic smuggling group. Source says that Joseph Pagano is the brother of John Pagano, partner of Joseph Petrizzo and the late Patsy Pagano who was killed several years ago.
>
> [An organized crime figure] was identified as having told associates in New York he had gone to New Mexico to arrange for the take over of carting firms in New Mexico and to control carting operations. [The organized crime figure mentioned] has a minor record as a gambling operator but is in direct partnership with Pagano in a number of other ventures. When Capone and Currieri bought out [the New Mexican firm]

> . . . they formed a new corporation called A1
> Carting. The same name used by Nicholas
> Ratteni in Westchester. Ratteni is Joseph
> Pagano's boss in the Genovese crime family.

The 1980 Annual Report by the Governor's Organized Crime Prevention Commission (Santa Fe, New Mexico) disclosed that they had recently begun hearings into the private carting industry which "found that in the early 1970s a number of individuals with close associates to New York City organized crime families became active in the private carting industry in New Mexico and that a pattern of violence, vandalism, equipment sabotage and business take-overs appears to have resulted."(24) In light of the allegations noted above, it is interesting that the Rio Rancho City Council approved the A-1 Carting Services application for continuing commercial refuse collection in the summer of 1982, according to Michael J. Hartranft in The Observer.(25)

In both the Florida and New Mexico summaries and reports, it is clear that toxic waste disposal problems were not contemplated. There is, therefore, no information uniquely related to toxics, although enough material is provided dealing with organized crime and the waste industry in general to provide a sober cautioning. And, it is reasonable to assume that organized crime companies in the waste industry would move into the very profitable toxic waste field in states outside of the Northeast.

Senators, I realize that the history and criminology I have presented is starkly pessimistic. However, the examples cited and the conclusions stated are minimal. I have not touched upon, for instance, the fuel oil industry which displays remarkably similar characteristics and, indeed, some of the same cast of characters, or the ancillary activities of many of the organized crime figures mentioned which run the gamut from industrial racketeering, narcotics, gambling, loan sharking, to murder. I have not mentioned other law enforcement figures who have had to conduct homicide investigations of individuals in the waste industry on their own time, using their own financial resources. The problems are acute and complex. Several years ago, I wrote a paper based on the theme that the development of leisure businesses in the 19th and 20th centuries presented professional criminals with their greatest opportunity for expansion. To find that they not only seized that opportunity, but have now penetrated closer to the center of modern industrial society by their forays into the petro-chemical industry is indeed disheartening.

* * * * * * * * * * * * * * * * * * * *

NOTES

1. Gore, 1980 House Hearing, 9.

2. Gore, 12.

3. Hearing, 6672.

4. Hearing, 7021.

5. Hearing, 7021.

6. Hearing, 7022.

7. Gore, Hearing, 16.

8. Ibid.

9. Ibid.

10. The New York Times, 25 September 1967.

11. The New York Times, 29 September 1974.

12. The New York Times, 18 October 1980.

13. Newark Star Ledger, 18 October 1980.

14. The New York Times, 18 October 1980.

15. State of New Jersey v. New Jersey Trade Waste Association, et. al., Indictment, 10.

16. Colton, 20 July 1980.

17. Attorneys for SCA Services, "Brief," 4 April 1979.

18. Minutes from meeting 30 June 1975.

19. Home News, 18 January 1983.

20. The New York Times, 11 December 1975.

21. Newsday, 15 September 1981.

22. The New York Times, 16 December 1980.

23. Ibid.

24. 1980 Annual Report by the Governor's Organized Crime Prevention Commission, 13.

25. The Observer, 5 August 1982.

Chapter 6
On the Need for the Waste Industry Disclosure Law

Published in Pennsylvania House Conservation Committee, <u>Hearing</u>, February 15, 1990.

I am a Professor in the Administration of Justice Department, The Pennsylvania State University. My research into the illicit disposal of waste began in 1980. In 1983 I testified on organized crime's involvement in both the garbage and toxic waste disposal industries before the U.S. Senate Permanent Subcommittee on Investigations. Now, seven years later, I thank the Pennsylvania House Conservation Committee for inviting me to speak on the proposed Waste Industry Disclosure Law.

In general, passage of this act accompanied by vigorous enforcement would go a long way in protecting the degradation of Pennsylvania's environment and the health of its citizens. This has become apparent to me in the course of my research particularly on the illicit disposal of toxic waste. There are firms with appalling histories of environmental violations which are today licensed by Pennsylvania's Department of Environmental Resources to transport hazardous wastes. It seems elementary to me that Pennsylvania should do all it can to keep these firms out of the Commonwealth. Enlightened regulation built on reliable and full information is surely important in stopping polluters. It is also obviously essential that firms and their major officers who have either serious criminal records, or are otherwise known to have committed serious crimes, not be licensed.

I wish to speak in some detail about two of these enterprises whose deeply troubling records are without doubt known to DER officials. Both firms are headquartered in South Kearny, New Jersey. The first is Sanitary Waste Carriers, Inc., which has Pennsylvania license number PA-AH0258. That permits the company to transport wastes that are ignitable, corrosive, reactive, ep toxic, and toxic both liquid and solid. The only category of waste that Sanitary is not permitted to carry in Pennsylvania is called "acute hazardous." The second firm which I will comment on is S & W Waste Inc., and its license, PA-AHO141, covers all categories.

It is probable that most Pennsylvanians are unfamiliar with Sanitary Waste Carriers. But I am certain that many are familiar with this company

105

under several different names. For the residents of Somerset County and others in Cambria, Clarion, Jefferson, Indiana, and Westmoreland counties, in south-central and western Pennsylvania, the firm Modern Transportation or Modern Resources or Modern Earthline will no doubt ring a bell. This company particularly under the name Modern Earthline was at the center of an environmental scandal and criminal conspiracy of quite massive proportions. When the City of Philadelphia was forced by the EPA to stop dumping its sewage into the Atlantic Ocean, it contracted with Modern Earthline to have it dumped in the above-mentioned counties as part of the strip mine reclamation effort. Sludge from Philadelphia was used to cover stripped land.

Modern Earthline was a complicated joint venture consisting at first of the company Modern Transportation located in Kearny, and a firm called Environmental Services which was itself a partnership between SCA Services of Passaic, Inc., a subsidiary of SCA Services, Inc., and Wastequid, Inc., a subsidiary of Scientific, Inc.   After the joint venture was formed, Environmental Services went through several name changes. In the fall of 1979 it was called the Earthline Company, hence the joint venture's best known appellation--Modern Earthline. There were subsequent name changes unnecessary for us to track. On the Modern side of the venture the head man was Richard J. Miele; on the Earthline side it was Anthony D. Gaess.

The sludge enterprise was expensive. Philadelphia paid Modern Earthline about $600,000 from 1977 through 1980 to develop a demonstration project to determine the feasibility of using sludge to revegetate stripped land. In addition, Philadelphia paid Modern Earthline $403,000 to provide equipment and equipment operators to transport the sludge; $831,000 to develop sites and apply the sludge; and $599,000 to transport sludge in those same years. Then in 1980 came the big money: several contracts were renewed for sums of $3,500,000 and $3,760,000. To secure these contracts, Modern Earthline and its major officers bribed its major competitor for the sludge contract. In addition, it swindled the City of Philadelphia through the use of fraudulent invoices and many other artifices in order to raise the money to pay the bribe. In 1984, Modern Earthline and its major officers were found guilty of violating federal fraud statutes.

Even earlier, however, both Gaess and Miele had been guilty of crimes which had quite serious potential environmental effects. Miele, for instance, had been charged by a federal Grand Jury in Newark, New Jersey, with obstruction of justice. This came about as a result of a plea bargain in connection with a $40,000 "kickback" scheme involving Kearny public officials and Modern Transportation's contract to haul from the Kearny Sewage Treatment plant. At that time there was also an investigation into Modern's alleged illegal disposal of hazardous waste through the Kearny sewer system. This investigation, DEP reported in an internal document, simply got lost in the shuffle of other matters. This report also notes that Richard Miele destroyed company records that probably showed the bribes paid to municipal officials. In any case, on April 25, 1982, Miele pled guilty to falsifying 1978-79 corporate tax records. He was fined and sentenced to three years probation.

Continuing on Miele, there was evidence collected by the New Jersey Division of Criminal Justice strongly suggesting that Modern Transportation's ocean sludge dumping operations during the 1970s acted to cover the illicit disposal of toxic wastes. Supposedly, Modern mixed chemical wastes with sludge at its South Amboy facility, placed this stuff on its barges, and then dumped it at sea. This allegation is especially disturbing given the vast amount of sludge Modern Earthline carted to western Pennsylvania. One must wonder whether or not that sludge (called mine mix which I have been informed was a combination of compost and almost raw sewage; residents in Somerset County report toilet paper, tampons, kotex pads, and other such items were visible in the mixture) was also laced with toxic chemical wastes. On this issue it is important to note that a former high functionary in New Jersey's DEP reported to New York State investigators as early as 1981 that Modern Transportation was "highly suspect of mishandling its wastes." And finally on this point of environmental hazards let me add that Miele owned the majority interest in a company which operated a New Jersey landfill closed by the DEP for environmental violations.

The major issue for Pennsylvania residents and members of this Committee to consider at this moment is whether the DER is serving the best interests of the Commonwealth in licensing Sanitary Waste Carriers. That corporation, as is well known, was originally listed in New Jersey files as the correct legal entity for Modern Transportation. According to New York investigators, the firm was incorporated on the 10th of April, 1972, in Hudson County. Around the mid-1980s, New Jersey's DEP noted that Sanitary Waste Carriers "holds DEP Registration number S-3560," and that it had become the trucking arm for Modern in its hauling businesses. In a DEP report under the heading of "Corporate Veils, Employment Contracts, and Other Questionable Business Practices," the following was said: "the conduct of any substantial part of Modern's business through Sanitary Waste Carriers raises a question which goes to the reliability of the operation in terms of financial responsibility to the public." The writer surmised the function of having a separate corporation for trucks was probably to limit Modern's liability in case of an accident. He added that having DEP allow operators to have arrangements which shielded their assets from damage claims, especially at a time when the state is trying to build confidence that the hazardous waste industry can operate responsibly, is bad policy.

As noted above, the leader of the Earthline half of the sludge operation was Anthony Gaess. For his part in the Philadelphia sludge conspiracy, Gaess pled guilty to mail fraud. Like his co-conspirator Miele, that was not Gaess's first problem with the law, however. While Federal

Judge Edward N. Cahn was considering Gaess's sentence, the New York State
Senate Select Committee on Crime detailed some of its information on
Gaess's activities in a letter. This was part of an unsuccessful effort to have
the judge pressure Gaess to reveal what he knew of "illegal toxic dumping"
from Maine to Maryland. The Committee said, for example, that Gaess had
been sued by the federal government in 1979 for toxic waste dumping at the
Kin-Buc landfill in New Jersey. But not all the relevant information on Kin-
Buc was revealed; Gaess could, it was believed, aid in identifying all the
generators and carters who dumped millions of gallons of carcinogenic waste.
Moreover, the New York Committee had information that Gaess had
established a toxic waste dump on the grounds of the Mount Airy Lodge,
Pocono Summit, Pa. In this particular endeavor, he was helped by an
associate who was murdered in 1980. Finally, the Committee informed the
judge, it had reliable information that Gaess illegally disposed of cylinders of
military poison gases, burying them somewhere in the vicinity of Middletown,
New York.

The second firm I wish to speak about is S & W Waste, Inc. It is also
headquartered in South Kearny, New Jersey. I believe S & W is important
for the Committee to consider because it too has a remarkably unenviable
record earned over the years in New Jersey, Pennsylvania and Ohio. In the
latter state, S & W was responsible for the deaths of three men in December
1984 when a municipal garbage incineration plant located in Akron, Ohio,
exploded. This tragic incident in which several other workers were severely
burned was caused by ongoing illegal toxic waste disposal.

This is how it happened. On December 19, 1984, George Walsh drove
his Peterbilt rig and forty-foot trailer into S & W's yard. His assignment was
to pick up a load of SAWOL for delivery the next day to a municipal garbage
incineration plant in Akron, Ohio. SAWOL is a mixture of sawdust and
waste oil which when burned with garbage increases the efficiency of the
combustion process. However, it wasn't SAWOL that was loaded onto
Walsh's trailer. It was instead a 45,000 pound mixture of sawdust and 2500
gallons of highly flammable toxic chemicals--xylene, toluene, and methyl ethyl
ketone (MEK). These are solvents used as prime ingredients in paint
thinners and they burn like gasoline: xylene's flash point is 63, toluene 40
and MEK 21.

The solvents came to S & W from a nearby Mobil paint manufacturing
plant. Once a year, Mobil pours these solvents into its paint storage tanks to
liquify the sludge coagulated at the tank bottoms. The tanks are then
pumped out by vacuum trucks whose drivers are specially trained in safety
procedures and equipped with breathing devices to avoid inhaling the highly
toxic fumes evaporating from the volatile solvents. The vacuum trucks on
December 19th brought the sludge and solvents to S & W where it was
dumped into pits and mixed with sawdust. There it was transferred to
Walsh's trailer with a frontend loader. Stone and gravel were also scooped

up with the sawdust. These became important clues in determining what caused the killer explosion that shattered the Akron garbage plant the next day right after the delivery of Walsh's load of solvents, sludge and sawdust.

Two loads of the Mobil tank sludge arrived at S & W on the nineteenth. The proper way to dispose of the solvent/sludge rock/gravel mixture was to let it stand in tanks for a few days until the liquids separated from the sludge. Then the solids should have been put into barrels for burning in a specially designed incinerator. S & W's yard man, Freddy Lee "Latin" Jones, did nothing of the sort. The mix was placed in an open pit with sawdust and then the concoction was given to Walsh to transport to Akron. The Ohio garbage plant could not burn solvents, however.

Walsh left S & W about nine in the evening. He drove through Pennsylvania on the turnpike heading west until two in the morning when he pulled off at a truckstop to sleep. The following day around noon he arrived at the Akron incinerator. Walsh dumped his load of toxic saturated sawdust for transmission onto a garbage conveyor belt that fed the fire. His mixture gave off the strong smell of nail polish remover. Several employees at the plant noticed the odor and that the mixture was off-color, gray, instead of the usual dark SAWOL. A driver in line behind Walsh smelling the odor came over to the conveyor belt took a deeper whiff and yelled, "Hey man, that's lacquer thinner. It could blow up." He dumped his load and quickly left.

As the sawdust solvent mix travelled through the plant on the belt, more plant workers noticed the strong odor and complained to the supervisor. They wanted Walsh's load rejected, remembering several earlier S & W shipments which had given them headaches, had made them dizzy. The supervisor called S & W and asked what was in Walsh's SAWOL. He was told it was harmless fragrances but that S & W would check and call back. The strong odor of nail polish remover lingered about the receiving platform and along the belt because the vapors given off by the solvents are two to three times heavier than air and don't dissipate. Concerned, the supervisor told his workers to take their lunch break until the odor left. That lunch break saved their lives.

It took around twenty minutes for the SAWOL to move along the conveyor belt to the top of a tower where it fell into a fire. When that happened, a flame raced up the column of SAWOL falling from the top of the tower causing an explosion that tore holes in the tower's concrete roof. This first explosion triggered fire alarms and safety devices which shut down the conveyor system and switched the boilers over to natural gas instead of the burning garbage and toxic SAWOL.

Two floor employees, Bell and Katona, ran over to where three outside workmen were repairing a garbage chute under the conveyor belt on the receiving room floor. Bell asked if they were alright. They were, but then as Bell and Katona turned to walk away, they heard a swooshing sound and were enveloped in flames. A second deadly explosion had raced across the receiving room.

Akron fire investigators were able to reconstruct these events: the first explosion created a vacuum with its ignition; air from the surrounding area rushed in to fill the vacuum sucking with it the explosive vapors emanating from the toxic SAWOL not yet on the conveyor. The whoosh heard by Bell and Katona was the air and vapors rocketing to fill the vacuum. A wall of flame was produced from the ignition of the vapors.

Bell's clothes melted and he crawled through flames to the control room. Katona's clothes caught fire but he was able to run to the control room where others beat out the flames. No one was yet safe. In the midst of this chaos a third explosion erupted which blew in the protective glass window of the control room.

Fire trucks rolled up to the plant and the search for survivors started. As firemen poked through the debris, they heard a siren-like sound which they thought at first was an alarm. It was actually the scream of the outside repairman who had been working on the receiving floor. He died a short time later. His two companions were found an hour or so later under the debris. They were dead; burned so badly they had become extraordinarily rigid making it very difficult to move them onto a litter. Seven others were injured and hospitalized.

During these appalling events, S & W was trying to call the Akron plant letting them know Walsh's load was safe. S & W couldn't get through with this reassuring message because the phone line was jammed with calls for emergency assistance.

The three dead men were autopsied, their blood samples sent to a special lab in Dallas for testing. Xylene, toluene and MEK were in the blood, meaning the men had inhaled the vapors before dying. Bell and Katona were hospitalized and treated for burns. Bell's hands were in particularly bad shape requiring much painful therapy before they could function again.

After unloading the solvent/sludge/sawdust/rock and gravel mix, Walsh left the Akron plant driving to the nearest truck stop where he steam cleaned his rig and went on to pick up a load of grain for the return trip. The cleaning wasn't very efficient and three days later, when Ohio investigators questioned Walsh in New Jersey his truck still smelled of paint thinner. This likely meant the grain picked up elements of the carcinogens xylene, toluene and MEK.

The Akron prosecutor's office assigned Fred Zuck one of their ablest attorneys to supervise the investigation after the initial inquiry uncovered the hazardous content of the SAWOL. Contact was made with the New Jersey Attorney General's office and its assistance requested. S & W responded on February 3, 1985, announcing that it had retained Edwin Stier the former director of the New Jersey Division of Criminal Justice to "conduct an objective inquiry and to advise the corporation and board of directors of the nature of the facts surrounding the incident."

When Stier was chief of criminal justice he headed up a federally-funded investigation of toxic waste dumping in New Jersey. At that time, Stier's Division reported that S & W was a high "priority" target for illegal dumping of toxic wastes. Surveillance of the South Kearny facility by State Police investigators and others from Stier's group indicated substantial dumping illegalities.

The hiring of Stier was reported in <u>The New York Times</u> and stimulated someone to send an anonymous letter to the Akron prosecutor's office. It correctly claimed that S & W had been caught dumping drums in Jersey City, an action investigated by the New Jersey Department of Environmental Protection. The letter writer also claimed that S & W was tied to organized crime figures prominent in the waste disposal business. Along with this information, the writer gave a short history of S & W's known involvement in illicit toxic waste disposal. This included around 1500 drums dumped at the McAdoo Landfill in Pennsylvania in 1978. This was nothing compared to the next bit of information which held that S & W along with others had dumped over 100,000 barrels in an abandoned strip mine in Old Forge, Pennsylvania. By chance a television documentary crew from ABC filming material for a documentary called "The Killing Grounds" had caught S & W in the act.

There was still more in the letter. The Akron prosecutor learned that one of S & W's employees involved in the deadly shipment had been charged with water pollution violations by the federal government while working for a different company. More importantly, another individual involved in the shipment from New Jersey to Akron had been convicted by federal authorities of toxic waste dumping and was sentenced to prison. At the time he arranged for S & W to deliver the SAWOL to Akron he was out on bail pending appeal.

The president of S & W, William Moscatello, had other legal problems. He was arrested in New Jersey for possession of stolen property and received two years probation. The letter writer strongly suggested Akron investigators check out this case. They did and found S & W had been caught with stolen truck trailers during a stake-out by New Jersey state troopers. This information came from of an investigation into the dumping of toxic paint sludge hauled away from a Ford Motor plant then in Mahwah,

New Jersey. An undercover witness for the FBI, Harold Kaufman, had testified that Ford's toxic waste disposal business was controlled by organized crime. S & W and a notorious dumping operation named Duane Marine had been designated as the firms to remove Ford's paint sludge by the racketeers.

The material in the letter was checked, confirmed and added up to this: S & W had been suspected of committing various major environmental crimes and was labelled a "priority" concern by Stier's special toxic waste unit; it had been filmed committing environmental crimes; it had been tied to organized crime by Kaufman who was New Jersey's foremost witness in waste disposal/mob investigations and prosecutions; and it worked with and employed convicted dumpers and polluters. But yet when all was said and done, S & W seemed to live a charmed life in New Jersey.

Several investigative findings convinced the Akron prosecutors that the deadly S & W shipment to Akron was no accident. When the vacuum truck operators who had removed the sludge from Mobil to S & W were questioned about how S & W had handled the previous tank bottoms, they stated they mixed them with sawdust in a pit creating SAWOL. Furthermore, visual inspection of S & W's yard revealed there were no open containers where the mixture of solvents and sludge could have been held if anyone had actually wanted to follow safe procedures. The containers would have allowed the sludge to settle and the solvents to have been decanted.

The vacuum truck operators also told the Akron investigators that the day after the explosion, they had brought the final load from Mobil to S & W. On this occasion, S & W attempted to handle the sludge and solvents in the right way but couldn't. Without the open container in the ground positioned for the vacuum truck to unload it was hopeless. Even when S & W had the truck attempt to pump the sludge out into a container level with it, they found it still couldn't be done. The inescapable conclusion was that S & W could only dispose of the Mobil tank bottoms illegally.

That wasn't all. Another line of inquiry disclosed more evidence that S & W's shipment to Akron was no accident. The company had also shipped hazardous wastes to two landfills in Ohio. Asked by prosecutor Zuck if they had ever rejected S & W loads, the landfill operators turned over a stack of rejection slips which showed S & W had shipped drums of toxics concealed in loads of solid waste and loads with dangerously low flash points.

The disposal of the solvents and sludge, Zuck believed, was clearly reckless. This laid S & W open to the charge of involuntary manslaughter because their reckless disposal of the explosive SAWOL in an incinerator could not have been a mistake. The only way S & W could dispose of the toxic Mobil waste was the wrong way. On April 12, 1985, an Akron Grand Jury charged S & W, William and Harry Moscatello and three other employees with involuntary manslaughter.

The charge of manslaughter was based on a new statute that had only become effective on October 1, 1984. Section 3734.99 of the Ohio Revised Code made the recklessly illegal transportation and disposal of hazardous waste a felony. If the death of a person was the proximate result of the commission of such a felony, then the charge of involuntary manslaughter could be brought. That was what Zuck did.

The trial began on October 15, 1985. The main defense team came from the Akron law firm of Blakemore, Rosen, Meeker and Varian, who represented S & W and the Moscatellos. They had been recommended to the defendants' New Jersey attorneys by U.S. Senator Howard Metzenbaum.

The prosecution's first witness testified that on December 14, 1985, six days before the fatal explosion, Mobil's officers instructed S & W on the proper procedures for the safe handling and disposal of the paint bottoms. S & W's representative was given a written profile of the mix which said it was flammable below 100 degrees and thus had to very carefully handled. S & W's man was told the mixture must be kept in an open top tank; the solids had to settle, the liquids decanted. The solids should then be mixed with cement dust for further solidification and sent in barrels to a secure landfill. When the vacuum trucks left Mobil, S & W was called by Mobil and told to have the open top container ready. Of course S & W had no such containers. Corroboration of these facts came from S & W's delivery supervisor. The first several witnesses therefore gave compelling testimony that S & W's handling of Mobil's waste was deliberately and recklessly illegal.

Over the following two weeks of the trial, Zuck called 41 witnesses and introduced 110 exhibits into evidence. Witness after Witness stated that S & W ignored clear instructions on the handling of the Mobil waste. Some also testified that S & W also clearly violated the terms of its agreement with the Akron garbage plant. Witnesses said that the explosion could only have been caused by the S & W load because stones and gravel found embedded in the area of the Akron plant after the explosion matched stones and gravel later found in S & W's South Kearny yard. Medical evidence was introduced that conclusively established the dead men had xylene, toluene and MEK in their blood at the time of death thus linking their deaths directly to the S & W load.

On the completion of the prosecution's case, the trial judge directed a verdict of acquittal on the ground the evidence was insufficient to prove recklessness. The Judge declared there was no evidence that the system of handling hazardous waste by S & W was inadequate from which one could infer recklessness. The Judge added that a single, accidental breach of a duty or care, no matter how tragic the result, without any other evidence of the act being done knowingly, purposely, or recklessly, could not rise to the level of criminal conduct.

The judicial verdict if it had been rendered in a civil case could have been reversed on appeal. However, directed verdicts of acquittal in criminal cases cannot be appealed because of the double jeopardy prohibition in the U.S. Constitution. The Judge's verdict foreclosed review by a higher court and effectively silenced any protest by the prosecution.

There was a coda to the Akron tragedy. Throughout the pre-indictment investigation and later during the trial, S & W's attorneys asserted that their insurance coverage was more than sufficient to compensate the victims of the explosion. This did not prove to be the case.

The civil suits to determine the damage due the victims resulted in awards totaling $8.5 million to the families of the three men killed: $1 million each to Bell and Katona for their burns, and from $65,000 to &150,000 to five other workers less severely injured. When it was time to pay, the victims learned the insurance carrier for S & W (as well as for the trucking company that employed Walsh, the City of Akron, and the operator of the garbage plant) was insolvent. Integrity insurance was its name.

S & W then argued it would be forced into bankruptcy if it had to pay its share of the damages. The Court relieved it of all but a small share of its liability. In the end, S & W put $600,000 into the pot from which the ten victims were paid. This despite its claim to be worth $10 million in 1984 and 1985. In fact, the damages supposedly paid by S & W were passed on to the taxpayers of New Jersey through the New Jersey Guarantee Association which liquidated the allegedly insolvent insurance carrier. The costs of liquidation were then passed on as surcharges to insurance premiums paid by New Jersey's citizens. The liquidation process placed a cap of $300,000 on the individual awards to the Akron victims from the insurance pool that was supposed to protect S & W and its trucker. The other defendants, the City of Akron and the plant operator, had their shares of the damages paid by the Ohio Insurance Guarantee Association although their liability was far less than S & W's.

The declination of responsibility for the damages on the grounds of its shaky financial condition raised a question as to S & W's ability to continue to operate in the toxic waste business. Anyone in the industry must be licensed and a criterion for licensing is the ability to pay damages in the event of accident or catastrophe. Nonetheless, S & W continues in business today.

Could an Akron-like explosion happen today somewhere else? In August 1988, the coordinator of the Solid/Hazardous Waste Background Investigation Unit of New Jersey testified before a congressional committee seeking an update on New Jersey's situation. He stressed two points: hazardous waste companies with extremely poor environmental track records are operating and most ominously hazardous waste companies consistently label hazardous waste as non-hazardous material in order to illegally increase profits by disposing of the material at less costly rates. He said, in effect, they're still doing it.

I strongly commend this committee's action and support this legislation. When it passes, I urge those responsible for its implementation to take a long and serious look at the firms I have mentioned along with others also currently licensed whose past records are marked by serious environmental violations, and compelling evidence of criminality.

PART IV:
THE STATE, ORGANIZED CRIME, AND POLITICAL MURDER

Chapter 7
Fascism, Organized Crime and Foreign Policy: The Assassination of Carlo Tresca

Published in Simon and Spitzer (eds.) <u>Research in Law, Deviance and Social Control</u> (JAI Press, 1982)

"On the evening of January 11, 1943, . . . Carlo Tresca, the last leader of the American anarchists,"(1) was murdered in New York City. The crime was never solved. After almost four decades an examination of this murder is in order. Whether or not Tresca was "the last leader of the American anarchists," as the conservative historian Francis Russell wrote, that he was a significant radical leader who was assassinated justifies a close inquiry. Moreover, Tresca's biography exemplifies the role of political violence in the twentieth century. Even the most minimal investigations of the man pulls one close to the murderous machinations of dictatorial and imperialistic states, a world of plots and counter-plots, and personal tragedy bounded by assassination.

The study of Tresca's murder leads to a consideration of political murder. Ever since the assassination of President Kennedy, political murder and violence has become something of a cottage industry for social scientists.(2) Without exception, however, the social scientific literature sheds neither theoretical nor empirical light on the Tresca murder. In fact, in probability the most ambitious collection of studies on the general subject of political assassination, Tresca is given almost no attention. Even his name is misspelled.(3) The essay in which Tresca (Treska) is mentioned, suggests that various immigrant groups in America "were a favorite recruiting and funding source, and frequently an organizational base, for national liberation and political revolutionary movements that were aimed primarily at the European homeland."(4) These remarks are very close to the truth, and yet they are misleading. In this formulation America is merely an inert mass in which the daffy peasants from abroad work out their problems of adjustment in typically violent ways. Such notions, naturally, leave out the contributions which American society had made to stability and violence both at home and abroad, as well as to the ways in which immigrant groups constructed their hyphenated identities.

What follows is a detailed inquiry into the facts and interpretations of Tresca's assassination which lead inevitably to a much wider context. There are four major sections: first a brief history of Tresca's life is presented in

117

order to indicate his historical significance; second, an analysis of the leading interpretation of the assassination which places responsibility on Communist agents; next, a long look at the second, clearly minor, claim that Fascists murdered Tresca; finally, the wider context is explored. Specifically this deals with the impact of World War II on perceptions of Communism and Fascism, on the reconstruction of Italy, and on the world of cooperative endeavor among professional criminals, Fascist politicians and certain American military and foreign policy officials. Before getting on with it, one last word. This is not a legal brief; it is not my intention to present a case that would satisfy, even remotely, judicial requirements. The type of evidence for that is lacking. My purpose is to write the history of a crime which reveals much about the relationships among organized crime, political movements and assassination.

PART I--CARLO TRESCA(5)

Carlo Tresca came to the United States in 1904 after already establishing his credentials as a political radical in Italy. Although he came from a moderately well-to-do family, he was early attracted to radical doctrines and participated in several demonstrations in northern Italy. At one point he fled Italy for Geneva where he made the acquaintance of among others, Benito Mussolini who was a radical socialist at this stage of his career. Tresca left Europe because of legal problems in Italy. He had been convicted of anarchist propaganda and, rather than face a prison sentence, emigrated.

Tresca rose to prominence in the United States as a member of the Industrial Workers of the World (IWW) an organization notorious for its anarcho-syndicalist doctrine, its "free-speech" battles especially in California and Washington, and extraordinary strikes at Lawrence, Massachusetts and Paterson, New Jersey. Probably no labor organization in American history called forth such a sustained attack by both State and Federal law agents as the IWW. And this for several reasons: (1) the rhetoric if not the ideology of the IWW was unabashedly revolutionary, and this during a period in which capitalism in general experienced world-wide challenges culminating in the Russian revolution; (2) the IWW sought to organize in one industrial union not only industrial workers but migrant workers and the "new immigrants" from southern and eastern Europe; (3) together these factors and others called forth a very strong "law and order" response. A demonology was created by both state propagandists and middle-class reformers which fueled both quantitative and qualitative legal changes. New laws were enacted at the state and federal levels which fought the IWW through conspiracy and sedition statutes, and, during World War I, with newly enacted espionage laws. Local, state and federal battles with the IWW involved, however, much more than law creation. Implementation of new law as well as the often artless but effective exercise of police power on all three levels (including extra-legal vigilantes) combined to defeat the IWW. In the 1920s, it was rendered totally ineffective.

Through this period of struggles and repression, Carlo Tresca played a significant but relatively unknown role. There are bits and pieces of his activities recorded in these battles, snippets of information acknowledging his importance especially in organizing Italian immigrant workers, but little more.

Tresca and the IWW parted company in 1917 for a variety of reasons mostly a result of the perpetual internal ideological and political struggles characteristic of American radical movements. His leaving the IWW, however, had little impact on his standing as a commanding figure in America's radical intelligentsia--that amalgam of literary types and working-class organizers which found expression and companionship in Mabel Doge Luhan's salon and other bohemian dens. Even though he was no longer a "Wobbly" Tresca still was the subject of various state and federal investigations whose primary purpose was to find some way to deport him. The years 1919-1921 were after all marked by what some have called "deportation deliriums" and the infamous Red Scare. Tresca was extremely well known as an anarchist with a profound influence in Italian communities across urban America, and was therefore, considered exceptionally dangerous. It was during the 1920s that Tresca turned to his favored vocation and gave full attention to Il Martello (The Hammer) an Italian-language newspaper which he published and edited in New York. As a radical polemicist, Tresca concentrated on the grievances of Italian-Americans--the depredations of professional criminals and racketeers in their communities and trades, the general exploitation of immigrants by capitalism and their salvation through allegiance to anarcho-syndicalist principles and organizations. In addition, Tresca began in the early 1920s to warn against a seductively attractive new political system which had emerged in Italy--Fascism. For Tresca, "gangsterism," capitalism, and now Fascism were all enemies to be ceaselessly fought.

While his activism was deeply embedded in the social conditions of New York's Italian communities, he maneuvered on a larger stage. He used Il Martello to advertise birth control. For this he was indicted, tried, convicted and sentenced to a year in the federal penitentiary in Atlanta, Georgia.(6) He assumed an important role in the Sacco-Vanzetti defense committee, being responsible for coordinating both local Italian demonstrations and European expressions of solidarity. Throughout the 1920s, Tresca was active in several spheres of radical life, concentrating on racketeers invading such trades as brick layings in New York, as well as the international reactions to the Sacco-Vanzetti affair and the rise of Fascism in Italy.

In the 1930s, Tresca continued his work as an anarchist writer and organizer, but as the world changed so did his concerns. The center stage of

Il Martello increasingly became the menace of Fascism which had itself changed over the course of the decade. Tresca worked as a political organizer in Italian communities and put forth radical alternatives to the burgeoning fascist organizations cropping up in a number of eastern cities.

At the same time two other "world" events occupied him and shaped his opposition to Communism. The first was the Spanish Civil War. Tresca was an implacable foe of Franco and worked diligently to prevent his triumph. He devoted much effort in recruiting volunteers to serve in the Abraham Lincoln brigade and promote the Republican cause. These concerns cast him into a close collaboration with American communists who were also active in the same endeavors. Tresca, of course, had been deeply involved with communists at many times in his career, but he had carefully distinguished his anarchist ideology from that of his more totalitarian colleagues and friends for many years. The rather careful relationships, the fine points of radical doctrine differentiating the many competing left-wing groups active in America for so many years, began to irrevocably fall apart in the mid-1930s. For Tresca, the Spanish Civil War provided the reasons for a final break with Communism. In that complex tragedy played out in Spain, both Fascism and Communism stood exposed as enemies. While Tresca had long been a foe of fascism, the GPU purges of the anarchists in the republican forces, especially the Barcelona and Catalan representatives, turned Tresca into a dedicated enemy of Communism. Moreover, the assassination of a number of his friends by the Soviet secret police deepened his hostility to Communism.

The Russian purge trials were also significant in shaping Tresca's development as a radical anti-communist. Acting with other radicals and/or reformers, Tresca helped form the Commission of Inquiry into the Charges Made Against Leon Trotsky in the Moscow Trials. The Commission, as part of their inquiry into the purges and Trotsky's alleged role in subversion, journeyed to Trotsky's exile in Mexico City. They conducted a hearing in order to allow Trotsky to defend himself.

By the late 1930s then, Tresca as far as the international scene was concerned, was an avowed enemy of Stalinist Communism, Italian and Nazi Fascism, and as always, Capitalism. On the local level, Tresca was privy to a rather amazing amount of information concerning the illegal activities of both Communists and Fascists in America, ranging from assassinations to spying. A good deal of his information found its way into Il Martello but not into the hands of his "capitalist" enemies. Increasingly Tresca's framework for "law and order" and is own safety was his cadre of followers recruited from New York's Italian communities.

With the dawn of World War II, Tresca's anti-Fascist evangelism seemed to pay off, and he was deeply involved in various Italian-American organizations designed to play some sort of role (hopefully on the lines of anarcho-syndicalism) in the post-war reconstruction of Italy. Through 1941 and 1942 his major task was to keep Communists (back in the fold since Hitler's attack on Russia) and Fascists out of these organizations. Communists may have returned to respectability with liberals after the debacle of the Nazi-Soviet non-aggression pact was obliterated, but Tresca always kept in mind their incredible treacheries during the 1930s culminating in the Trotsky assassination.

PART II--THE ASSASSINATION: COMMUNIST PLOT?

In general such was the scope and significance of Tresca's activities until World War II. Let us now turn to the assassination and the primary interpretive focus--Communist responsibility. The secondary sources tell an amazing story. Benjamin Gitlow, ex-Communist and cold war evangel, wrote one version in his memoirs published in 1948.(7) Carlo Tresca's enemy was "totalitarianism" in any form, wrote Gitlow, but by chance he "became involved with the ugliest form of the police state, the OGPU." This began with an "affair with a communist sculptress" whose studio was some sort of meeting place for OGPU agents. Drawn in first by the woman, Tresca soon allegedly "became enmeshed" in OGPU matters.(8) The Spanish Civil War destroyed this supposedly burgeoning partnership and in particular set Tresca against the OGPU assassin Vittorio Vidali, alias Carlos Contreras, alias Enea Sormenti. This enmity heated up considerably in 1942 when the Office of War Information (OWI) helped to organize the Italian-American Victory Council which was to aid in the eventual occupation of Italy by allied forces. Tresca was utterly opposed to Communists belonging to the Council. Gitlow wrote: "The Mazzini Society, the leading organization of Italian anti-fascists, due to Tresca's pressure, had previously, in principle, adopted the policy of excluding all totalitarian and fascist elements from the society."(9) Having successfully kept both Communists and Fascists out of the Mazzini Society, Tresca next worked to exclude them from the Italian-American Victory Council. The Communist response to Tresca's implacable stance was murder, Gitlow argued.

> Tresca personally told me, shortly before he was murdered, that he knew that the OGPU assassin Sormenti, . . . was in New York. . . . But he boldly bragged that before Sormenti could get him, his boys would get Sormenti. A few weeks before he was murdered, Tresca confided to friends that he had seen Sormenti in New York and cryptically remarked: "Where he is I smell murder. I wonder who will be the next victim."(10)

As far as Gitlow is concerned, the assassin was Sormenti, Joseph Stalin's surrogate. But Gitlow's explanation, on the face of it, is neither satisfying nor convincing. The reliability of reformed Communists and radicals writing during the Cold War on such sensitive issues is suspect.

Moreover, there appears nothing to substantiate that the OGPU assassin Vidali-Contreras-Sormenti murdered Tresca.   The conversation about "smelling death" while chilling, proves nothing about the murder.   This particular claim, however, has had a remarkable tenacity and was somewhat more fully developed in another essay.

Francis Russell the Sacco-Vanzetti scholar mentioned in the opening of this essay, wrote in 1964 that Tresca and Vidali had known each other in New York during the 1920s. For reasons unmentioned, Vidali was deported to Mexico in 1928, and there "developed his talent for political assassination by arranging the murder of Antonio Mella, a Cuban ex-communist who had turned against the party and was living in exile."(11) Vidali next appeared in Spain acting as the OGPU's chief assassin during the Spanish Civil War. Russell explained that "It was by his arrangement, and in spite of the protests of the Spanish Republican Prime Minister Juan Negrin, that Andres Nin was executed."(12) With Franco's victory, Vidali was forced to escape the country and made his way back to Mexico.   While there he concentrated on murdering exiled Leon Trotsky although he did not succeed. Perhaps it was Vidali's plotting in Mexico which brought him back to Tresca's attention, for in May, 1942, Tresca "denounced him on the front page of Il Martello as a 'commandant of spies, thieves and assassins.'"(13)   But consider Russell's words as he related the murder: "He [Vidali] was the type who liked to take his revenge personally.   Possibly he was the gunman waiting near the Fifth Avenue entrance to Tresca's office building. Almost certainly he was in the getaway car."(14) "Possibly" and "almost certainly" are not words inspired to resolve doubt and remove uncertainty.

There is little doubt that Vidali was an OGPU (NKVD) assassin, but so far nothing but hyperbole and wild conjecture connect him to Tresca's murder. Before abandoning this theme, however, it is fruitful to discuss the fullest exposition of the argument. This appeared in a pamphlet published by an organization known as the Tresca Memorial Committee. Shortly after Tresca's death, several of his most eminent friends and associates formed the Committee with the intent of pressuring various police agencies to vigorously pursue the case. The Chairman of the Committee was the venerable Socialist Norman Thomas. Other Committee members included Angelica Balabanoff, William Henry Chamberlin, John Dewey, John Haynes Holmes, Sidney Hook, A. Philip Randolph, Oswald Garrison Villard, Bertram D. Wolfe, and Edmund Wilson. In October 1945 the Committee published Who Killed Carlo Tresca? which summarized what they knew about the killing, and ended with some suggestions for stirring "the authorities out of their lethargy in the Tresca situation."

HELP GET JUSTICE
FOR CARLO TRESCA

To stir the authorities out of their lethargy in the Tresca situation, we need the aid of every man and woman who recognizes the menace in unpunished political murder.

Here are things you can do to help:

1.   *District Attorney Frank S. Hogan is running for re-election in November on both the Democratic and Republican tickets. Write him at once at 155 Leonard Street, New York City, and tell him he owes it to the whole community to demonstrate clearly before election day that he is actually doing something definite toward apprehending Carlo Tresca's slayers.*

2.   *Urge him to invite the Federal Bureau of Investigation to investigate the international aspects of this case, in view of Tresca's relentless attacks upon totalitarian groups and individuals connected with them.*

3.   *Write to the newly appointed Police Commissioner, Arthur W. Wallander, Police Headquarters, New York City. Urge him to make a new and independent investigation of the Tresca mystery, re-examining closely every clue in hand and every person known to have knowledge of Tresca's political conflicts.*

4.   *If you have information that might shed light on the identity of Tresca's slayers, and indicate the specific reason why he was killed, tell this committee or the District Attorney, or both. One of the guarantees in the $5,000 reward offer mentioned earlier is that such information will be received in absolute confidence.*

TRESCA MEMORIAL COMMITTEE

In the Committee's summary of Tresca's relations with Communists they noted that as early as 1933 he became "alienated" by the growth of the "Stalinist bureaucracy" and Communist efforts "to rule or wreck the American labor movement."(15)   In the following year, Tresca broke with the Communists because of their disruption of a strike in the New York hotel industry. The Committee then discussed Vidali-Sormenti-Contreras and the Soviet Secret Police. Tresca "charged the GPU with the murders of Camillo Berneri, Anarchist leader, in Barcelona; of Rudolph Klement, a lieutenant of Leon Trotsky, in Paris; of Andreas Nin, in the Spanish Civil War; of Trotsky himself, in Mexico; of Ignatz Reiss, a former GPU agent who had made a political break with the stalinists; and of others struck down in Spain and France."(16) Next, they claim that Tresca singled out Vidali in an article in Il Martello in May, 1942, accusing him of moving against the Mazzini Society "by order of Stalin."(17) This accusation was repeated in another article in the same month.

In the summer of 1942, verbal confrontations between Tresca and Communists increased. During the June convention of the Mazzini Society, Tresca "read a letter from Mexico City signed with the name of Vittorio Vidali (Sormenti) and addressed to Pietro Allegra, writer for the Stalinist organ, L'Unita del Popolo."(18) According to Tresca, the intent of the letter was to urge that "Randolfo Pacciardi, who had commanded the Garibaldi Brigade in Spain," move to unify with Communists in a common Anti-Fascist front. The Committee states: "'Unity is spoken of here," said Tresca, "as it was spoken of during the Spanish Civil War. . . . You, Pacciardi, and I will be put in jail or killed with a shot in the back of the head. If the Communists were sincere I would extend my hand for common action. But they are not, and they must not enter the Mazzini Society.'"(19)

In continuing their summary, the Tresca Memorial Committee pointed out that the Mazzini Society publicly claimed that the "files of the Daily Worker are replete with invective against Tresca." Moreover, Tresca was verbally attacked by the Italian National Commission of the Communist Party and by the Italian Communist Pietro Allegra mentioned above who wrote a pamphlet titled The Moral Suicide of Carlo Tresca. The pamphlet and indeed most of the Daily Worker stories commented on by the Mazzini Society appeared in 1938. Nevertheless, the Committee strongly implied that the Communists' invective was a call for Tresca's assassination. For instance, the Committee quoted a statement written by the Italian National Commission of the Communist Party and published in 1938 in a foreign language "Stalinist" paper in New York which called for "'Tresca's isolation'" as a necessary step for all anti-Fascists. In addition, the article holds that "'Without any other preoccupation except that of protecting and safeguarding anti-Fascism, we therefore launch a fraternal appeal to the militants of all groups or political parties, . . . that in the common interest they make Tresca understand that police informers will no longer be tolerated in the political and labor movement. . .'"(20)

The conclusion drawn by the Memorial Committee that the written statements in the Communist press were thinly veiled calls for the elimination of Tresca was apparently shared by Tresca himself. The Committee quoted a Tresca statement which appeared in Il Martello:

> If Tresca is alive, sane, and not inclined to die,
> either physically or politically, to [please] the
> melancholy Pietrino [Allegra] . . . it will be
> necessary to put him out of the way, definitely.

In a word: what is needed is a George Mink, a
member of the Communist Party of America, and
the murderer of our comrades Berneri and
Barieri. It must not be said that such a
sanguinary and macabre idea has not come into
the heads of the four lazy fellows who make up
the Italian National Commission of the CPA. . .
. However, they would like to urge some
firebrand to put the idea into execution. It is not
always necessary to arm the hand of a killer;
enough to inflame the mind of a fanatic.(21)

This is not an apologetic for the Stalinist State or its secret police. So
many fine writers have shown that assassinations, indeed mass murder; were
part of its logic and activities.(22) For instance, Hannah Arendt in The
Origins of Totalitarianism commented on the institution of the secret police
in what she called "the only two authentic forms of totalitarian domination"
in this century: The dictatorship of National Socialism after 1938, and the
dictatorship of Bolshevism since 1930."(23) In both instances "the power
nucleus of the country" were the "superefficient and supercompetent services
of the secret police." Their tasks were "the abolition of the distinction
between a foreign country and a home country, between foreign and
domestic affairs."(24) For those segments of the secret police which operated
outside the home country, Arendt remarked, acted as "transmission belts
which constantly transform the ostensibly foreign policy of the totalitarian
state into the potentially domestic business of the totalitarian movement."(25)
To repeat, this is no defense for totalitarianism and such organizations as the
OGPU (NKVD). Nevertheless, it is clear that there is no basis in the
available evidence for the claim of Communist responsibility for this
particular assassination, or that Vidali had anything to do with Tresca's
murder.

Having said this, however, begs the question of why Vidali and the
Communists have been so tenaciously selected by so many writers as the
killers? The answer most likely lies in the fact that many of the memoirs of
radical life in America during the first four decades of the twentieth century
are "confessions" in which the perfidy of Communism is the major motif.(26)
Authors like Benjamin Gitlow who knew Tresca or wrote about his
assassination placed the murder within the Communist orbit as another
example of Communist treachery. The murder of the undoubtedly anti-
Communist Tresca must have been carried out by Vidali and the Soviet
Secret Police because such actions were a logical part of the nightmare of
Communist subversion. The necessity for such formulations flowed from re-
reformed, re-born, ex-radicals as a significant feature of their penance and
was eagerly embraced in the first flush of the Cold War. Indeed, "The Great
Fear" of Communist subversion, as David Caute termed it,(27) which had
only been temporarily and tenuously shelved during World War II,

demanded, at least implicitly, that all political murders could and should be traced to Communism.

The deep-seated fear of Communist subversion explains the development and strength of the Vidali story, and also underlines other issues illuminated by studying the Tresca assassination. It forms part of the reason why Fascism was not taken very seriously as the force or movement behind the Tresca assassination, except by dedicated anti-Fascists.   This became all the more obvious in 1944 and after, when it seemed to some that Fascism had been defeated forever, and that Communism was emerging from the War as the West's most potent enemy. The military defeat of Fascism and National Socialism destroyed the Fascist movements, and thereby, ironically enough, led to the almost immediate rehabilitation of individual Fascists who could help in countering Communism. What seemed to concern policy makers were large-scale movements not that particular political identities of individuals especially their past political allegiances. Cooperation with Fascists was by the latter years of World War II not the same at all as cooperation with Fascism. It is well known that in foreign policy matters there was a decided shift in the identification of the enemy. Once it was clear to both America and Great Britain that Germany and Italy were going to lose the war, planning shifted toward the containment of Communism meaning both Soviet imperialism and what is now called Euro-communism.

PART III--THE ASSASSINATION: FASCIST PLOT?

The argument of Fascist responsibility in the assassination was never developed in the literature. But the Tresca Memorial Committee explored it, and naturally enough, it formed a constant refrain in such Communist vehicles as the Daily Worker. The discussion of Fascist culpability revolves around the activities of Generoso Pope publisher of the "authoritative and influential Il Progresso Italo-Americano of New York the largest such daily in the nation, with a paid circulation of 81,000,"(28) as well as the daily Corriere d'America which was discontinued in 1942.   In pursuing the question of Fascist responsibility it is important to discuss first the Fascist presence in America, and then Pope's involvement in Fascist activities. Following this, evidence of Pope's involvement in the murder is presented. Proceeding in this way allows discussion of the broader issues surrounding the assassination.

As a way of initially placing the Fascist presence in America, consider Mussolini's reception in the United States. According to John P. Diggins' history, Mussolini enjoys a "vast popularity" which was a "product of the press."(29)   Diggins pointed out that the The New York Times correspondents' writing on Italy approved of Fascism and Mussolini. One of the most prolific was Anne O'Hare McCormick who "rhapsodized upon the feats of the Blackshirts and consistently defended the twists and turns of Mussolini's diplomacy, justifying the Ethiopian invasion, the Italian 'volunteers' in Spain, and the Rome-Berlin axis."(30) Like the extraordinarily influential The New York Times which featured so many rhapsodic articles on Fascism, the mass circulation Saturday Evening Post which had about three million subscribers in 1930 "effectively created a respectable image of Mussolini."(31) Indeed, Post writers did much more than make Mussolini respectable, they described him in numerous articles as a "political savior," and "economic genius," and the leader in the struggle of "virtue over vice."(32) In 1928, the Post went beyond description of Fascism and Mussolini, and in serial form published Mussolini's "autobiography." Negotiations for this publishing coup were carried out by the American Ambassador to Italy, Richard Washburn Child who was "infatuated with the Italian dictator and frequently conferred with him on the state of American opinion."(33) Most likely the Mussolini autobiography published in the Post was in fact written by Child with the aid of Mussolini's brother.(34) In addition to the mass circulation dailies and weeklies which tended to extol both the dictator and the movement out of deep conservative biases, several "former muckrakers were just as complimentary toward the Italian dictatorship." Among this group were Lincoln Steffens, S.S. McClure, and Ida Tarbell who after returning from Italy in 1926 "advised the State Department that the new labor laws in Italy constituted an admirable social experiment.(35)

The American adulation of Mussolini was not simply the product of the malign combination of romanticism and conservatism exemplified by Anne O'Hare McCormick of the Times and Ambassador Child, and it should be added William Randolph Hearst and his general manager, Colonel Frank Knox, who apparently idealized Mussolini.(36) In 1927, for instance, a press service was organized in the United States by the Italian Government to more forcefully proselytize the achievements of Mussolini and Fascism. What is most interesting about this development, however, is that the idea originated with "Thomas W. Lamont and Martin Egan, head of the J.P. Morgan Company's press department."(37) The Italian Government's role was kept secret and Edgar Sisson who had been the managing editor of Collier's magazine was hired to oversee the operation.(38)

One might expect that the early 1930s would have witnessed a growing sense of disquietude with dictators and the burgeoning European movement of Fascism as the Great Depression intensified and German National Socialism grew ever more truculent and powerful. But with some notable exceptions, Mussolini was perceived during those years as a "pioneering economist and peacekeeping diplomat."(39) It was necessary for some American Fascist sympathizers to distinguish in print the differences between the Nazi State and Italy. They succeeded so well that "Ultimately the growing menace of Nazism, far from discrediting Mussolini, enabled the

Italian Premier to parade as the enlightened statesman who could counter Hitler's aggressive designs."(40)

The Infrastructure of Italian-American Fascism

Mussolini and Italian Fascism enjoyed for almost two decades considerable support if not adulation in the American press. But the most important vehicle of Fascist propaganda in the United States was the Italian-language press which consisted of between 130 and 150 publications from 1920 to 1940.(41) Early on, most of America's Italian papers received their news of Italian developments from "the Stefani news agency, a pro-Fascist establishment at the time of the March on Rome which would assume enormous importance after Mussolini consolidated his power."(42) There was, of course, an anti-Fascist press exemplified in many ways by Tresca's Il Martello. But in terms of money and circulation it was overwhelmed. Diggins noted that in New York, "the readership of the pro-Fascist papers outnumbered the opposition about ten to one."(43) The primary advertisers in the Fascist papers were bankers, manufacturers, large restaurants and wineries. The power structure in Italian-American communities was the major supporters of the Fascist press.(44) The largest circulation of any Italian-language newspaper was enjoyed by Pope's Il Progresso Italo-Americano. And it is incontrovertible that the Italian government played a major role in Pope's purchase of the paper. In October 1928, Giacomo de Martino, Italy's Ambassador to Washington, wrote to the Consul General of Italy that "it is imperative that ownership remain in the hands of someone absolutely faithful to Italy."(45) Generoso Pope, who was considered more than safe, then purchased the paper with financial backing from several Italian-American bankers.(46)

The Fascist presence in Italian-American communities cannot be properly gauged or appreciated by understanding only the press. Beyond propaganda, there was an institutional nexus of Fascist organizations operating in America. Even before Mussolini's March on Rome, a number of influential Italian-Americans had advocated a "Fascio" in America. They were joined in their enthusiasm by Fascist party officials in Italy. Opposed to this expansion, at least initially, were "career diplomats in the Italian Foreign Office. . ."(47) The various groups attempted to pressure Mussolini who displayed a certain amount of caution over the issue or probability more accurately the methodology of Fascist penetration in America. The "Imperial dream" was, nevertheless, "close to Il Duce's heart," and in 1924 Count Ignazio Thaon di Revel was sent to America to manage the Fascist League of North America.(48) The FLNA had been created before 1924 by Italian-Americans but it suffered from severe organizational problems. So, in 1924, the Italian Fascist Party sent their agent to "reactivate" it. The success of Count di Revel was rapid, and "local affiliations began to spring up throughout the country numbering at least seventy branches with a dues paying membership of between 6,000 and 7,000." The avowed purpose of the

FLNA and affiliated organizations was to battle subversion meaning Communism, atheism, and class hatred.(49) The FLNA was directed by the Bureau of Fascism Abroad a section of "one of the most important departments of the government, the Ministry of Popular Culture."(50) According to Article 1 of the Bureau's statute, affiliated groups were "'unions of Italians residing abroad who have adopted for their private and civic life obedience to Il Duce and to the laws of Fascism . . .'"(51)

The growth of the FLNA was finally halted in the late 1920s when it was revealed that a special "court of discipline" had been established within the organization.(52) In addition, there had been growing criticism of the FLNA on the part of organized anti-Fascists, "complaints that officials were putting pressure on residents not to take out naturalization papers," reports of reprisals against the relatives of Italian-American businessmen and of discrimination against their commercial contacts "if they did not cooperate fully with Italian consular officials."(53) Criticism in Congress led by Senator William E. Borah of Idaho, increased until there was a call for a Congressional investigation. Count di Revel moved to head off this problem. On December 22, 1929, the Fascist League of North America was dissolved.(54)

As far as American Fascism was concerned, the disappearance of the FLNA meant little. Another organization almost immediately replaced it. Formed by Domenico Trombetta publisher of Il Grido della Stirpe, probably the most rabidly Fascist Italian-language paper in America, was the Lictor Federation which was described as "'the direct successor of the old Fascist League of North America.'"(55) Affiliated with the Lictor Federation were other organizations such as the Italian Historical Society, the Federation of Italian World War Veterans, the Dante Alighieri Society which was organized as a literary society in 1890 and embraced Fascism in the early 1920s, the Casa Italiana of Columbia University, the Italian Disabled Soldiers Society, Afterwork (Dopolavoro) Clubs, and the Order of the Sons of Italy which claimed about 200,000 members.(56) The last group mentioned was split over the issue of Fascism with important defections occurring as early as 1922 when one of the officers were to Italy "to place at the feet of Mussolini the allegiance of the entire society."(57) Most significantly, in New York important members of the Sons of Italy such as Fiorello La Guardia and State Senator Cotillo left the parent body and formed a "separate fraternity." However, as Diggins wrote, this initial dissension allowed the Sons of Italy to purge the recalcitrant, and the organization went on to propagate Fascist propaganda "with many of its chapters acting as unofficial organs of the Italian Ministry of Popular Culture."(58)

The first line of Fascist organization in American communities was a series of local Fascist clubs some affiliated with the Sons of Italy and the Lictor Federation and some so evanescent that affiliation never took place. The identity of many of these clubs comes from the Italian-American Fascist press and from those clubs which participated in large-scale Fascist demonstrations primarily in the New York Metropolitan area during the Ethiopian War, 1935-36. The uncovering of Fascist clubs at this time was the work of Gaetano Salvemini an Italian anti-Fascist exile who did more to counter Fascist propaganda on a world-wide scale than probably anyone else.(59) Salvemini's impeccable research revealed the communal level of Fascist organization in the New York area. He identified the officers and activities of numerous Fascist clubs active in the 1930s. The clubs discussed include the following: Abraham Lincoln Club, The Bronx, which mixed and merged in a short period of time with the Mario Sonzini Fascio, and finally surfaced as the Benito Mussolini Club, The Bronx; Francesco Crispi, Manhattan; Maria Jose, Manhattan; Association of Italians Abroad, Manhattan; Edmondo Rossoni, Brooklyn; Araldo di Crollalanza, Manhattan; Isole Maggio Del Bronx, The Bronx; Diciotto Novembre 1935, XIV, which translates to Eighteenth of November 1935, Fourteenth Year of the Fascist Era, Brooklyn; Flatbush Italo-American Association, Brooklyn; Armando Diaz, Ozone Park, Long Island; Camillo Benso di Cavour, Brooklyn; and many others.(60)

Each of these clubs participated in an extraordinary open-air demonstration which was held at Morristown, New Jersey, on August 16, 1936. This rally was held on the grounds of the religious organization Maestre Pie Filippine nuns who managed a number of Italian schools throughout the United States. In addition to the Fascist clubs mentioned above, others from Tarrytown, Peekskill, Yonkers, Staten Island, and Newburgh took part in the event. Most of the participants, both men and women, marched past the Italian Consul General wearing uniforms, especially black shirts.(61)

Generoso Pope--Fascist

Fascism was well established and deeply entrenched within the associational life of Italian-Americans. And, Generoso Pope as publisher of pro-Fascist papers was clearly one of the most important Fascist propagandists. Pope's Fascist activities, however, were not entirely subsumed by his newspapers. As a man of influence Pope played a role in legitimizing Fascism by his participation in public events that extolled Mussolini and Italian Fascism. For instance, he was a member of a committee which arranged for the reception of Italo Balbo, Italian airman, on January 3, 1929. A few weeks later, Piero Parini, "director of the bureau of Italians abroad and director general of the Italian schools abroad at the Rome Foreign Office" came to the United States. While in New York, Parini received the "honorary title of deputy sheriff of New York County" at a dinner arranged for him by Pope.(62) The Chief Editor of Il Progresso Italo-Americano was Italo Carlo Falbo who was a "friend" of Mussolini's and also represented the state-controlled Stefani News Agency in New York. Pope engaged in fund raising activities for the Reverend Joseph Congedo's educational endeavors among Italian-American children which were laced with Fascist propaganda. The Reverend Congedo was knighted by Mussolini in 1932 for spreading the Fascist gospel.(63)   Early in 1934, Pope was a featured speaker at, what Salvemini described, "the golden anniversary of the priesthood of the Fascist Reverend Francis P. Grassi, paster of Saint Anthony, Wakefield, Bronx."(64) In March 1934, Pope sent a representative to a "birthday of Fascism" party held at the Hotel Ambassador which had been promoted by <u>Grido della Stirpe</u> and was attended by Count di Revel and other Fascists. In October 1934 Pope sent a telegram of "greeting and approval" to the Lictor Association which had "promoted a celebration of the March on Rome."(65) Similar demonstrations took place in 1935 with Pope either speaking or as one of the distinguished guests. He was especially prominent in defending "Italy's right to civilize Ethiopa" during that year.(66) In March 1936, Pope was a "guest of honor" at the by now annual Birthday of Fascism celebration. Salvemini described this particular event as follows:

> Pope congratulated Trombetta upon the magnificent struggle for "Italianism" he had valiantly led for years. He recorded that <u>Il Grido della Stirpe</u> had been the first paper ever to print his picture, many years before. He observed that the 1936 celebration of the Birthday of Fascism coincided with exceptionally important events which were revealing to the world the material and moral strength of Mussolini's Italy and the gallantry of our Army gloriously fighting on the battlefields of East Africa. . . .(67)

When Italy achieved the battlefield victory in Ethiopa which was celebrated on June 13, 1936, at Madison Square Garden, Pope gave to the Italian consul a check for $100,000 collected through his newspapers. Pope's aid to Italy was recognized by Mussolini who telegrammed effusive thanks to Pope.(68)

With only one substantial exception which will be dealt with shortly, there has never been any question of Pope's background in the Italian-American Fascist movement. There is, however, a substantial question dealing with Pope and Fascism which is concerned with precisely when he actually denounced Mussolini and Fascism.   Furthermore, there are interpretive problems dealing with the meaning of Pope's anti-Fascist statements coming as they did in general after Italy and America were at war.

Let us first deal with the exception--Pope's defender who claimed that he was never even a Fascist sympathizer.   In the Spring of 1941, Congressman Samuel Dickstein from New York took up the cudgels for Pope in the House of Representatives and answered newspaper allegations that

Pope had been a Fascist sympathizer. Dickstein acknowledged that Pope was the publisher of two, as he put it, "outstanding Italian newspapers," and then added that he was a benefactor to the poor and that "He always condemned fascism and the Mussolini movement."(69) Several months later, Dickstein continued his defense in Congress. This was necessary because in the interim, Dickstein's original statement had sparked a flurry of mail to Congress which as Dickstein stated accused "Mr. Pope of fascistic activities" and criticized the Congressman. The main perpetrator of this mail, Dickstein charged, was Giralomo Valenti, editor of <u>La Parola</u> and friend and associate of Carlo Tresca characterized as an "avowed provocateur."(70) Dickstein's defense in this case was to point out that Valenti, Tresca, and others were both Communist propagandists and secret Fascists. In particular, Dickstein singled out the Mazzini Society and stated that "the truth of the matter is, that the organizers of this society are Fascists who have fallen apart recently from the councils of the party, and who wish to propagate their own brand of Fascism against that officially adopted by the existing Fascist organization."(71)

Pope's defender, Samuel Dickstein, served in Congress from 1923 to 1944. His importance in Congressional, indeed American, history resolves around his key role in the formation of the House Committee of Un-American Activities. Dickstein was primarily concerned with Nazi propaganda and activities in America, although running a close second was anti-Communism. In 1934, Dickstein moved a resolution in Congress for a Committee to investigate Nazi activities and subversive propaganda. The resolution was authorized and John McCormick was selected as Chair. Three years later, Walter Goodman wrote, "Dickstein, whose inquisitorial passions had been choked rather than slaked by the McCormick hearings, introduced a resolution in language that opened up spectacular new investigative vistas. Now he wanted to probe all organizations 'found operating in the United States for the purpose of diffusing within the United States of slanderous or libelous un-American propaganda of religious, racial or subversive political prejudices which tends to incite the use of force and violence . . . whether such propaganda appears to be of foreign or domestic origin.'"(72)

Dickstein's defense of Pope could be construed as the actions of a man blinded by a total absorption in anti-Nazism and anti-Communism. But there is a more sinister possibility. According to a State Department document in the National Archives, there is reason to suspect that Dickstein and Pope were themselves conspirators. The document was written by Casimir P. Palmer who apparently worked as an investigator for the Department of State which received the communication on October 30, 1942. The substance of the report concerns a conversation between Palmer and his informant, a Mr. Lee. Palmer wrote:

The most important part of our conversation was this: Mr. Lee said that several months ago Richard Rollins instructed him to go to the Mazzini Society, at 1775 Broadway, and see the head of the organization, Professor Max Ascoli, who is also the director of the School of Social Research. "Dick" Rollins had instructed him (Lee) to offer the Prof. a sum of $5,000, if he would stop attacking the fascists and Mussolini; if the Prof. would stop attacking the fascists and Mussolini; if the Prof. would accept this offer Gereroso Pope, the publisher of two Italian newspapers here, would accept him, Prof. Ascoli, as a partner and guarantee him his job for life.

When Mr. Lee called at the headquarters of the Mazzini Society he was met by an old man who said he was in charge of the office and when he heard Mr. Lee's proposition "he flew into a rage," and Mr. Lee had no alternative but to leave the office.

From the talks Mr. Lee had with "Dick" Rollins it was clear that the proposition was made by Generoso Pope to Congressman Samuel Dickstein, who, in turn, communicated the offer to his "confidential secretary" Dick Rollins, who then instructed investigator Lee to submit the proposition to the head of the Mazzini Society.(73)

There is no corroboration for this alleged conspiracy which would indicate, if true, that Dickstein's motive in defending Pope was more than simple stupidity and blindness. In any case, it is clear that Pope's involvement in Fascist activities was still strong as late as the end of March 1941. In another State Department document which primarily concerned Domenico Trombetta "classified as a dangerous Italian agent, and the leading Fascist propagandist in America," a bit more of the Fascist web was unravelled. The report stated that Trombetta is "closely associated with Generoso POPE, Grand Officer of the Crown of Italy," and "influential member of the Democratic Party of New York City," and a member of one of the Appeals Boards of the Selective Service.(74) In dealing with Trombetta's finances the report claimed that he "probably receives financial support for his newspaper from Italian government sources, through POPE." Pope's political power in New York was traced to "one Jack Ingegnieros" who was president of the Federation of Italian-American Democratic Clubs in New York State, and who "recently formed the National Committee of the Italian-American Democratic Organization." Ingegnieros was employed by Pope on the staff of Il Progresso Italo-Americano. Finally it was noted that "in the event of hostilities between the United States Government and Italy, that Pope will [not] openly aid Italy, but he will continue his activities through Trombetta and Ingegnieros."(75)

War, of course, did come and forced Pope into a volte-face. The timing of his public reverse, its significance and sincerity are unclear and disputatious. What is certain is that sometime after March 29, 1941, probably in May, Pope had a meting with President Roosevelt who convinced "the Tammany publisher that Fascism, even Italian Fascism, was no longer respectable."(76) Following this meeting Il Progresso's editorial content and policy changed, and some mild criticism of Mussolini appeared. It is also true that this about face was quickly noted in Italy and that the Italian Ambassador to Washington wired Rome that "Pope was basically a political opportunist and therefore unreliable." But, even if Pope was overwhelmingly an opportunist and not an ideologue, it still appears far-fetched to conclude, as Diggins did, that Generoso Pope was a "much maligned man." Diggins maintained that Pope was not "a hopeless reactionary" because he stood "against Italian anti-Semitism" and supported Roosevelt and was loyal to the United States when war came between Italy and America.(77) The criteria suggested by Diggins in his appraisal are simply insufficient to move Pope into the class of the maligned.

Pope and the Assassination

We have discussed the first two points: the Fascist milieu in America, and Pope's Fascist activities. Now, let us turn to Pope's relationships with professional criminals, especially those involved in the murder of Tresca. All the extant evidence in the case points to Carmine Galante as the assassin.(78) Galante, who was himself gunned down in the summer of 1979 at which time he was reputed to be one of America's most important professional criminals, was in the early 1940s associated with a wide number of infamous criminals. Among the men he worked both with and for were racketeers Frank Garofalo, Joe and Peter Di Palermo, Joseph Bonanno, Tony Bender, John Dioguardi, Vito Genovese, Joseph Profaci, Vincenzo Martinez, and Frank Cetrano alias Chick Wilson, according to FBI files secured under the Freedom of Information Act, and the Tresca homicide files in the New York County District Attorney's Office.

Galante's location at the crime is clear. On the night when Tresca was murdered, he had been waiting in his office with Giuseppe Calabi, "an attorney and refugee from Italy, for four other men"(79) to work out details for a new anti-Fascist section of the Mazzini Society. The other men were Vanni Montana, secretary to Luigi Antonini who was president of the Italian-American Labour Council, a high official in the International Ladies Garment Workers Union, and one of the founders of the American Labor Party; John Sala, "an organizer for the Amalgamated Clothing Workers of America; Giovanni Profenna; and Gian Mario Lanzilotti."(80)

None of the invited men appeared and Tresca and Calabi left the office. They were on their way to a tavern when Tresca was shot from behind. The killer then jumped into a waiting car, which the police located later that evening. Two days later Galante was arrested. He had been seen entering the car about an hour before the murder. Galante who was on parole had left his parole officer that evening and been followed. Sidney Gross, his parole officer, "had learned that Galante was again associating with known criminals, and therefore had assigned two parole officers, Fred Berson and George Talianoff, to follow him."(81)

There is in government files and several research libraries an unsigned eight-page document which advances the most complete and sinister scenario of Pope's complicity in the murder committed by Galante.(82) The probable major author is Girolamo Valenti, friend and associate of Tresca and member of the Italian-American Victory Council. The reported stated that Carmine Galante was a member of the "Castellammare branch of the Italo American underworld" in New York, and that the two leaders of this branch were Frank Garofalo and Joseph Bonanno. Garofalo was also, it claimed, the "factotum of Generoso Pope," on his payroll acting as Pope's bodyguard and general "strong arm man." Starting in 1934, Pope using Garofalo began a campaign of intimidation against other Italian-language newspapers which were anti-Fascist and anti-Pope. Girolamo Valenti (see above) editor of La Stampa Libera was reportedly stopped by Garofalo "and told first rather subtly, then . . . very bluntly, that the personal propaganda against Generoso Pope was to cease and if those attacks were not discontinued dire consequences would result." Other papers involved in this campaign were Tresca's Il Martello, Frank Giordano's La Tribuna, and Marziale Sisca's La Follia. Giordano sold his paper and was then "placed on the staff" of Pope's Il Progresso; Siscal resisted and was beaten having his skull fractured. Tresca was not intimidated, and excoriated Pope and his gangster associates in Il Martello on October 28, 1934.

Animosity between Pope and Tresca continued throughout the decade. The situation became critical sometime in September 1942. This time it involved Pope, Garofalo, Tresca and a woman named Dolores Facconte who was an Assistant United States Attorney, and the "admitted and open paramour" of Garofalo. Around the tenth of September a dinner was arranged for Italian-Americans for the purpose of selling war bonds. Present were Pope, Facconte and Tresca. During the course of the evening Garofalo entered. Tresca rose up and loudly protested Garofalo's presence. He allegedly "threatened to attack Garofalo" in Il Martello. What followed was a verbally violent confrontation between Tresca and Garofalo with Tresca threatening to "expose Garofalo and to show the connection that existed between him and Pope." Garofalo's reported reply was "that before that would happen Tresca would be found dead in the gutter." In the days immediately following, Dolores Facconte visited Tresca several times imploring him not to expose her relationship with Garofalo. Tresca supposedly told her that he would not write about her affair, but that he intended "to crucify not only the 'mafia leader Garofalo, but also his boss Pope.'" The assumption flowing from this report is that Tresca was murdered in the interests of Garofalo and Pope by Galante a member of Garofalo's gang.

A number of FBI documents repeat the same story in substance. In one document dated January 13, 1943, the events at the dinner in September are recounted, and then it is noted that on the day after the murder the New York Office of the FBI was informed by teletype that a "source" suspects Frank Garofalo because of past quarrels and Tresca's imminent activities in the Mazzini Society. It is also clear from this document and others that Tresca had been an FBI informant and was last contacted by a Bureau Agent on January 6, 1943. A Confidential report on the case was sent by Special Agent E.E. Conroy to J. Edgar Hoover on February 5, 1943, which repeated the story but then added a most interesting dimension. It stated that the enmity between Tresca, Garofalo and Pope, was compounded by the efforts of Luigi Antonini, the renowned labor leader of the ILGWU and the American Labor Party, who had secretly formed an alliance with Pope. The alliance had been negotiated, it claimed, by Garofalo and one of Antonini's strong arm men. Its purpose was to legitimize Pope and thereby secure him a position on the Italian-American Victory Council. The alleged reasons for this deal included Pope's apparently secret contributions to an ILGWU sanitarium built in Los Angeles, and Pope's considerable political power in the Democratic Party. Another FBI document dated October 16, 1944, stated that a confidential informant "advised that the only thing that had kept Luigi Antonini from openly joining forces with Generoso Pope, was Carlo Tresca, and once Tresca was out of the picture there was no one to oppose Antonini's desire."

In the District Attorney's files there are a documents indicating that Galante worked for Garofalo and Joseph Bonanno, and one which added that members of the American Labor Party contributed funds for Galante's defense. Memoranda dated January 7, 1944, January 21, 1944, and August 18, 1944, in the same file state that Galante, Frank Cetrano, Tony Garappa, and Joe and Peter Di Palermo were paid $9,000 for the Tresca murder. The money was passed along by Joseph Parisi who by 1944 was associated with the International Brotherhood of Teamsters Local 27.

Carmine Galante was held for about eight months while the murder investigation proceeded. He was never indicted; in fact, no one was ever indicted for the murder and therefore there is little corroboration for the evidence gathered by the New York authorities or the FBI. While extraordinarily suggestive, the particular details of the assassination cannot be established. Nevertheless, the connections between Pope and Garofalo are not in doubt, nor are those between Garofalo and Galante. Furthermore, there is a memorandum in the D.A.'s files which provides another bond between Pope and the racketeers already mentioned.

Dated April 16, 1943, it noted that Vincenzo Martinez who was a "reporter for Il Progresso" was a close friend of both Garofalo and Galante and is himself "alleged to be a leader of the Mafia." Martinez was also an important Fascist official.   In 1927 when the Fascist League of North America established its notorious Court of Discipline, mentioned previously, Martinez was appointed one of the judges.(83)   The following year he became a member of the Fascist Central Council, the executive body of the FLNA.   Gaetano Salvemini recounted the Fascist organizations which Martinez directed and then pointed out that "On December 21, 1940, . . . he was indicted and then released on $1,000 bail under charges of extortion in his capacity as secretary of the Macaroni Employees Association, 225 Lafayette Street, New York City."(84)   Martinez is another palpable link connecting Pope to Garofalo and Galante, and moreover the social worlds of Italian-American Fascism and Italian-American organized crime.

There is still more evidence proving these links.   The address of Martinez's racketeer association turns out to be revealing.   There were other Fascist and criminal organizations located at 225 Lafayette Street.   The Italian-language weekly La Settimana which "unconditionally upheld the policy of Mussolini in the Ethiopian war" moved its office there in 1936.(85) The owner of the paper, Edward Corsi, who was a notorious Fascist sympathizer, according to Salvemini, was also reported to have attended a meeting during which the murder of Tresca was discussed.(86)   And, Baldo Aquilano "director of the press and public relations bureau" of the Sons of Italy, who had been one of the "directors of the Dante Alighieri Society," who founded the Italian Tourist Institute in 1927, and was a tireless propagandist for Italian Fascism had an office in the same building.(87)   The 1935 celebration of Italian Fascism conducted by the Association of Italians Abroad took place at 225 Lafayette Street.(88)

Besides Martinez, other racketeers worked out of the building.   A garment trucking association named the Five Borough Truckmen's Association headed by Dominick Didato alias Dick Terry, John Dioguardi, and Dioguardi's uncle, James Plumeri alias Jimmy Doyle was based there.(89)   In the summer of 1933, Didato was murdered and Plumeri wounded in a shoot-out in the office.(90)   Also quartered there was the Albert Marinelli Association, a political clubhouse.   In the mid-1930s, Marinelli was the County Clerk of Manhattan "and leader of the 2nd A.D. [Assembly District] who, . . . stands as the dominating influence in two assembly districts, the most powerful leader below 14th Street and the most powerful leader in Tammany Hall."(91)   According to material in the Thomas E. Dewey papers, Marinelli was a member of one of New York's most powerful criminal combinations which included Lucky Luciano, Vito Genovese, John Torrio, Vincent Mangano, Ciro Terranova, and Tony Bender.   Marinelli specialized in political fixing, voter intimidation and fraud, and numbers gambling.   Marinelli had also been associated with Dominick Didato in the late 1920s, and in the early 1930s was instrumental in aiding

Dioguardi, Plumeri and Didato take over the trucking industry in downtown Manhattan.(92)

Salvemini described 225 Lafayette as the "beehive" of Fascist activities. In the same sense it functioned as a "beehive" of Italian-American organized crime. There is no current evidence which places Pope there, but it cannot be denied that Vincenzo Martinez was more than able to represent Pope's interests whether Fascist or criminal. It is also certain that Pope was well acquainted with Marinelli. And finally, there is evidence that places Pope's reported "factotum," Garofalo, at 225 Lafayette on a regular basis. On the top floor was a restaurant, and Garofalo was a frequent patron.

The murder of Carlo Tresca was the malign product of the fundament of organized crime and Fascism. The killer was Carmine Galante, although it is not certain whether he murdered Tresca solely to please Garofalo, or to satisfy the mutual interests of Garofalo-Pope, or those more complex ones of Pope-Antonini, or most expansively the shared concerns of a considerably larger network of Fascists and professional criminals. Carlo Tresca had been a considerable problem for so many of the above for such a very long time, and Galante was tied to so many of those with murderous motives, that it is at this time impossible to get any closer to the murder.

PART IV--FASCISM, ORGANIZED CRIME AND FOREIGN POLICY

Although the main argument in the literature concerned with the murder of Carlo Tresca argues for Communist responsibility, it indicates much about the manner in which anti-Communism writes history and little about the murder. Clearly, the path to follow lies in the history of Italian-American Fascism and political violence. It is a movement and relationship difficult to track because, for one thing, Congress chose to ignore it. The 1930 investigation headed by Representative Hamilton Fish "worked for six months" on Communism and "totally ignored" Fascism.(93) The 1934 investigation chaired by John McCormack concentrated on "Communist and Nazi subversion and propaganda" but did not "examine similar activities by agents and supporters of Italian Fascism." it was much the same story when Congress revived the Un-American Activities Committees in 1938 under the leadership of Martin Dies of Texas.(94) The presence of Fascism and Fascists in Italian-American communities did not interest the Congress. This meant that those opposed to the movement would carry out their activities far from the mainstream of American life, and in many instances would meet with little more than cold hostility from investigative bodies. Their resort then, was to political violence as a counter to the perceived menace and as a result of their isolation.

Because Italian-American Fascism was allowed to prosper, so also was the concord between the political movement and Italian-American organized crime. This relationship which need not have been based on ideological affinities grew from a mutuality of interests which fostered cooperation. Some Italian-American criminals needed continuing access to Italy. Frank Garofalo, for example visited Italy in 1929, 1932, 1937 and 1938. The reasons why are not reported, but one of his allegedly legitimate businesses was the wholesale distribution of Italian cheese. Joseph Bonanno, reportedly Garofalo's major partner, maintained a strong connection with Sicily both before and after World War II. In fact, highly confidential information states that Bonanno made "frequent" trips to Italy, and in 1968 "was tried in Italy [we presume in absentia], for running narcotics and currency rackets." Other Bonanno and Garofalo associates deeply involved in both Italian and American organized crime activities include Gaetano Russo and Santo Sorge.

Also, Vito Genovese lived in Italy from 1937 to 1945 when he was returned to New York. Genovese's career is extremely interesting as it touches on so many of the substantive points of this inquiry. He was an infamous professional criminal; he had strong attachments to Italian-American Fascism, and to the Fascist Party in Italy; he was associated with Bonanno, Garofalo, and Galante; and was a suspect in the Tresca assassination. The allegation of his involvement is dated April 2, 1946, and stated that Tresca tried to prevent Genovese from starting a Maritime Club for Italian Seamen in New York during the 1930s. Tresca's opposition was based on Genovese's Fascist connections. When Genovese went to Italy his Fascist associates were deeply disturbed by Tresca's activities. The rest of the allegation suggested that Genovese sent back word to New York to have Tresca killed.(95)

For Fascist officials in Italy and America the cooperation of Italian-American criminals was important because of their social standing in Italian-American neighborhoods, and their very close ties to municipal politics at least. Professional criminals were very useful when it came to terrorizing or eliminating anti-Fascists. Professional criminals had financial and political influence, as well as a near monopoly on force and violence within immigrant communities. They were, therefore, significant for those interested in exporting Fascism. The two factions when not actually merged in particular individuals enjoyed a complex exchange relationship bonded by political and ethnic interests and loyalties.

The Impact of War

With the coming of World War II, this world of cooperative endeavor became somewhat precarious. But ironically, when American military planners decided to follow the wishes of the British and fight a Mediterranean War, professional criminals found new and powerful patrons. That the American commitment to fight in Italy flowed from British pressures had been cogently developed by Gabriel Kolko, and is an important issue to understand. There was, Kolko wrote, almost unanimous agreement among American political and military planners that "they did not wish to fight a war in the Mediterranean area . . ."(96) Nevertheless until the late spring of 1944 it was America's main battleground in Europe. This came about despite American opinion in the earlier stages of the war, that the British were unwilling to "open a second front in Europe" and that the Mediterranean was in the "British political sphere of influence."(97) Part of the tension which surrounded this issue came from the realization of "what these concessions to British strategy . . . could mean to long-term relations with the U.S.S.R."(98) What they meant, of course, was a guarantee of the Cold War--the fracturing of the post-war world into mutually antagonistic camps.

And yet Kolko argued, "it could hardly have been different" because the "fear in the West of the Left and Bolshevism was beyond negotiation . . ." Therefore, the "specific conditions that emerged" in Italy, for instance, "only reinforced the intensely deep-seated suspicions that the three Great Powers had of one another."(99) Italy "between 1943 and the end of the war . . . was a perfect microcosm of the European political situation as fascists, monarchists, liberals, leftists, and opportunists vied for power in endless Byzantine intrigues."(100) This was compounded by almost endless problems among the British and Americans as they vied with one another over which political, military, social and economic policies to pursue. And then the Russians added to "the burgeoning maze by entering the scene directly and through the agency of the Italian Communist party."(101)

Within this crazy quilt of factions and intrigue, there was still a "basic premise" to American "occupational planning" which "was to maintain existing governmental structures, laws, and even official political objectives."(102) In effect, this meant that the "status quo" would be preserved as much as possible. American and British planners, fearful of a chaotic situation because it would allow the Italian Communist party viewed as an extension of the Soviet Union to become the most powerful political force in Italy, "relied on existing fascist institutions, and save in the most blatant cases, fascist personnel."(103) American and British policy were both mutually hostile to what they perceived as Soviet imperialism in Italy, and its surrogate the Italian Communist party.

American and British policy disagreed, however, over which nation should play the major role in deciding post-war Italy's fundamental orientations--toward Britain or the United States. In preparing for the Yalta conference, the "basic assumption of British policy" that Italy was and should remain in the British sphere of influence was rejected by the U.S. State Department. Furthermore, it was positively stated that the United States has a "very real interest in the development of normal and mutually profitable trade relations, in the protection of American property and investments in Italy and in insuring that Italy becomes a positive force for peace and cooperation in the post-war world."(104) The bottom line, then, was that

Italy would be part of the American sphere of influence and that tremendous energy would be expended to prevent the Italian Left from taking power. While Britain and America could argue, even tenaciously over which Western power would control Italy, they were always in agreement that Communism must be thwarted.

For professional criminals both in America and Italy, the war and the occupation in Italy created the ground for new relationships, while doing as little as possible to upset the old. Even in the very beginning of American planning for the invasion of Sicily, there were tasks for professional criminals to perform. Naval intelligence enrolled criminals such as Joseph Lanza and Lucky Luciano in the cause.(105)

And then the OSS, forerunner of the CIA, availed itself of the services of professional criminals. American intelligence officers "maintained active contacts with native resistance groups, including the 'Mafia,'" amusingly characterized by the OSS as a "Sicilian separatist movement."(106) The networks of professional criminals called the Mafia were not a separatist movement, as American intelligence well knew. Nevertheless, American OSS operatives spent a great deal of time and energy "developing liaison with local political forces such as the . . . Mafia."(107) That American professional criminals had substantial common interests with the so-called Mafia was clear. That their interests merged with those of the invading and then occupying forces is also certain. In holding back the perceived Communist menace in Italy, intelligence organizations and their public partners in the military government utilized the services of professional criminals.

The case of Vito Genovese mentioned earlier who found employment with the Allied Military Government as an interpreter is instructive. He was apparently hired by Lieutenant Colonel Charles Poletti, commander of the Allied Military Government, and former New York State Supreme Court Justice and Lieutenant Governor of New York State. While supposedly acting as an interpreter, Genovese was deeply involved in black marketeering activities. Poletti whose political positions in New York were based on his power in the city's Italian-American communities, obviously knew of Genovese's criminal activities in New York during the 1930s. Also apparent was Genovese's membership in Italian Fascist organizations all during his self-imposed exile in Italy. Nevertheless, his rehabilitation in Italy through the good offices of the Allied Military Government was rapid. When Genovese was arrested in Italy by a zealous very junior officers in the CID, neither the Allied Military Government, nor the New York authorities who had in 1944 indicted Genovese for murder were interested in pursuing the case.(108) Intelligence officers were more concerned with the possible political ramifications of Genovese's arrest than anything else. Brigadier General Carter W. Clarke of American Military Intelligence noted the extreme sensitivity of the Genovese affair in a confidential memorandum sent to G-2 on June 30, 1945. Clark's recommendation to G-2 was that the "file" on Genovese was so "hot," that were it not for the fact that "At some later date" someone with knowledge of the case would "talk," he would "recommend that it be filed and no action taken."(109) The Genovese case is illustrative of the general proposition that professional criminals, American and Italian, continued to ply their trade for new patrons.

It should be obvious by now, that organizations such as the Italian-American Victory Council were of crucial importance in Italian-American affairs. Those Italian-Americans able to secure positions on such bodies would potentially have a great deal to say about the reconstruction of Italy. The battles for control of them are to be understood within a much wider context than simply intramural squabbles between various Italian-American factions. In the new world of the American imperium brought about by World War II, they were centrally located. This explains to some degree, the vehemence involved in the struggles for control by Tresca and Pope. It must be noted here that Italian-American labor leaders who were anti-Communist such as Luigi Antonini had their own important roles to play in reconstruction policies. Primarily, they were to operate as a wedge separating off militant trade unionists and Communists from the Italian labor movement. They exemplified the manner in which labor and capital could and should co-exist; they were the preferred alternative to revolutionary trade unionism.(110) War and reconstruction in Italy pulled together an almost bewildering amalgam of Fascists, criminals, and conservative trade unionists all working fundamentally to implement American foreign policy.

Carlo Tresca and his enemies played for very high stakes. On that cold night in January when Tresca lost his life, it seemed there was a world to win. But the fundament of organized crime and Fascism which caused his death, was already being engulfed by the larger concerns of military intelligence and foreign policy. The political murder of Tresca marks an important point in American political history.

* * * * * * * * * * * * * * * * * * *

NOTES

1. Francis Russell, (1964) <u>The Great Interlude: Neglected Events and Persons From the First World War to the Depression</u> (New York), 121.

2. See, for example, William J. Crotty (ed.), (1971), <u>Assassinations and the Political Order</u> (New York).

3. Ibid, 361.

4. Joseph Bensman, "Social and Institutional Factors Determining the Level of Violence and Political Assassination in the Operation of Society: A Theoretical Discussion," in Ibid; 360-61.

5. The secondary sources utilized in this brief biography fall into two sections:

1. Texts and biographies mentioning Tresca

Daniel Bell (1967) Marxian Socialism in the United States (Princeton); Philip S. Foner (1965) History of the Labor Movement in the United States, Vol. IV, The Industrial Workers of the World, 1905-1917 (New York); Walter goodman (1969) The Committee: The Extraordinary Career of the House Committee on Un-American Activities (Baltimore); Richard Drinnon (1961) Rebel in Paradise: A Biography of Emma Goldman (Boston); Isaac Deutscher (1963) The Prophet Outcast Vol. III, Trotsky: 1929-1940 (New York); Melvyn Dubofsky (1969) We Shall Be All: A History of the Industrial Workers of the World (Chicago); William L. O'Neill (1978) The Last Romantic: A Life of Max Eastman (New York); Robert A. Rosenstone (1975) Romantic Revolutionary: A Biography of John Reed (New York); W.A. Swanberg (1976) Norman Thomas: The Last Idealist (New York); Theodore Draper (1957) The Roots of American Communism (New York); Joyce L. Kornbluh, er., (1965) Rebel Voices: An I.W.W. Anthology (ann Arbor); Paul Arrich (1978) An American Anarchist: The Life of Volairine de Cleyre (Princeton, New Jersey); Robert Conquest (1968) The Great Terror: Stalin's Purge of the Thirties (New York); Francis Russell (1962) Tragedy in Dedham: The Story of the Sacco-Vanzetti Case (New York); Ray Ginger (1949) Eugene v. Debs: A Biography (New York); Henry David (1936) The History of the Haymarket Affair (New York).

2. Law and Radicalism

Lawrence Veysey, ed., (1970) Law and Resistance: American Attitudes Toward Authority (New York); Corliss Lamont, ed., (1968) The Trial of Elizabeth Gurley Flynn by the American Civil Liberties Union (New York); Louis f. Post (1923) The Deportations Delirium of Nineteen-Tenty: A Personal Narrative of an Historic Official Experience (Chicago); Robert K. Murray (1955) Red Scare: A Study in National Hysteria, 1919-1920 (Minneapolis); Constantin M. Panunzio (1921) The Deportation Cases of 1919-1920 (New York); Zechariah Chafee, Jr. (1941) Free Speech in the United States (Cambridge, Mass.); William Preston, Jr. (1963) Aliens and Dissenters: Federal Suppression of Radicals, 1903-1933 (Cambridge, MA); H.C. Peterson and Gilbert C. Fite (1957) Opponents of War (Madison, WI); Eldridge Foster Dowell (1939) A History of Criminal Syndicalism Legislation in the United States in the Johns Hopkins University Studies in Historical and Political Science, Vol. LVII (Baltimore); Homer Cummings and Carl McFarland (1937) Federal Justice: Chapters in the History of Justice and the Federal Executive (New York); J. Jaffee (1971) Crusade Against Radicalism:

New York During the Red Scare (New York); Robert Justin Goldstein (1978) "An American Gulag?" Summary Arrest and Emergency Detention of Political Dissidents in the United States," Columbia Human Rights Law Review, Vol. 10.

6. Tresca's arrest, trial and imprisonment were motivated by his anti-Fascist writing and the desire by the Italian government to silence him as FBI files clearly indicate.

7. Benjamin Gitlow (1948) The Whole of Their Lives: Communism in America--A Personal History and Intimate Portrayal of Its Leaders (New York).

8. Ibid, 338.

9. Ibid, 340.

10. Ibid, 340-41. The story of Vidali and the spanish Civil War has been covered by Deutscher, op cit. and Deutscher (1949) Stalin: A Political Biography (New York); Conquest, op cit.; and Hugh Thomas (1961) The Spanish Civil War (New York).

        According to Robert conquest, Vidali was attached to a special division of the K.K.V.D. (The Soviet Secret Police organization which had subsumed the OGPU in 1934) whose mission was assassinations outside the Soviet Union. A fair number of their assassinations were of members of the N.K.V.D. itself which was busy purging old members as well as "deviationists." In Spain the killing squads worked for the suppression of Spanish Trotskyism and the liquidation of key members of the Catalan Marxist party known by its initials--P.O.U.M. The reasons behind these seemingly contradictory killings in Spain are briefly described by Isaac Deutscher in his biography of Stalin. "The contradictions in which STalin involved himself led him to conduct from the Kremlin a civil war within the Spanish Civil War. The extreme Spanish Anarchist and Anarcho-Syndicalists fretted at the non-revolutionary tactics of the Communists. In Catalonia a semi-Trotskyist party, the P.O.U.M., tried to bring more social radicalism into the struggle. Stalin undertook the suppression of these un-orthodox elements on the left. He made their elimination from the republic's administration a condition of the sale of Soviet munitions to its Government. He dispatched to Spain, together with military instructors, agents of his political police, experts at heresy-hunting and purging, who established their own region of terror in the ranks of the republicans." The point of all this, writes Deutscher, was to "preserve for the Spanish Popular Front its republican respectability and to avoid antagonizing the British and French Governments."

11. Russell, The Great Interlude, 140.

12. Ibid.

13. Ibid., 140-141.

14. Ibid., 141.

15. Tresca Memorial Committee (1945) Who Killed Carlo Tresca ? (New York), 9.

16. Ibid., 11.

17. Ibid.

18. Ibid.

19. Ibid.

20. Ibid., 12-13.

21. Ibid., 13.

22. See the work of Conquest and Deutscher already cited.

23. Hannah Arendt (1958) The Origins of Totalitarianism (Cleveland), 419.

24. Ibid., 420.

25. Ibid., 421.

26. See Whittaker Chambers (1953) Witness (London); Hede Messing (1951) This Deception (New York); Herbert L. Packer (1962) Ex-Communist Witnesses (Standford); John Stratchey (1952) "The Absolutists" in The Nation, (October 4), 291-293; and Arthur Koestler (1948) Darkness at Noon (New York).

27. David Caute (1978) The Great Fear: The Anti-Communist Purge Under Truman and Eisenhower (New York).

28. Morris Schonbach (1958) "Native Fascism During the 1930s and 1940s (Ph.D. dissertation, UCLA), 105.

29. John P. Diggins (1972) Mussolini and Fascism: The View From America (Princeton), 24.

30. Ibid., 25.

31. Ibid., 27.

32. Ibid.

33. Ibid.

34. Ibid.

35. Ibid., 29.

36. Ibid., 48-49.

37. Ibid., 49-50.

38. Ibid. Diggins commented later in his book that the Morgan Company's interest was stimulated by the profit potential in managing Italy's $2,407,677,500 World War I debt, 152-153. Further discussion of the Bank of Morgan and Thomas Lamont is in Philip V. Cannistraro's Introduction to Gaetano Salvemini (1977) <u>Italian Fascist Activities in the United States</u> (New York), xv.

39. Diggins, 37.

40. Ibid., 40.

41. Ibid., 81-82 (note 6).

42. Ibid., 81.

43. Ibid., 83.

44. Ibid., 83-84.

45. Ibid., 85.

46. Ibid. Concerning Generoso Pope, Luigi Barzini later to be a world-famous writer, then the publisher of <u>Il Corriere</u> wrote to Mussolini that "Pope was an unreliable opportunist associated with Tammany Hall and the Mafia," 86.

47. Ibid., 88.

48. Ibid., 90.

49. Ibid., 91.

50. Schonbach, 74.

51. Ibid., 75.

52. Diggins, 93. On this subject Salvemini discussed and quoted an article by Marcus Duffield which stated that "the Court was designed after the fashion

of the special military tribune in Italy, for the purpose of preserving rigid discipline among the Fascists here and to meter out exemplary punishments ... Suspension or expulsion from the Fascist League is the commonest form of punishment imposed by these courts; but for more grave offenses there are penalties of corresponding severity. Confiscation of property, revocation of Italian citizenship and boycott lie in store for him who turns anti-Fascist, for he is a traitor. The League ipso facto penalizes any of its members who become naturalized citizens by taking away their rights and privileges as members of the Fascist organization in Italy. They may constitute a deprivation of no mean significance, for without the magic membership card, Italo-Americans traveling in Italy or transacting business there, are likely to find their way beset by innumerable difficulties," 72.

53. Schonbach, 80.

54. Ibid., 81-82; Diggins, 93.

55. Schonbach, 83; Diggins, 94.

56. Schonbach, 82-84; Diggins, 94-99.

57. Diggins, 94-95.

58. Ibid.

59. See Philip Cannistraro's Introduction to Salvemini.

60. Salvemini, 215-242.

61. Ibid., 209-214.

62. Ibid., 68.

63. Ibid., 153.

64. Ibid., 175.

65. Ibid., 178.

66. Ibid., 205.

67. Ibid., 206-207.

68. Ibid., 208.

69. Congressional Record, House of Representatives, 25 March 1941, 2569.

70. Congressional Record, House of Representatives, 17 June 1941, 5279.

71. Ibid., 5280.

72. Walter Goodman (1969) The Committee: The Extraordinary Career of the House Committee on Un-American Activities (Baltimore), 13-14.

73. U.S. Department of State, 740 EW.00119 Control (Italy portion of the Central Decimal File of the Department of State) in the National Archives, Diplomatic Branch, #865.20211 Mazzini Society.

74. State Department Document, National Archives, #800.011811 Trombetta, Domenico/4.

75. Ibid.

76. Diggins, 348.

77. Ibid., 348-49.

78. The extant evidence comes from FBI files on Tresca, about 1,000 pages, and the New York City investigation of the murder, files in the District Attorney of New York County Office. Access to both sets were obtained under the provisions of the Freedom of Information Act.

79. Tresca Memorial Committee, 7.

80. Ibid.

81. Ibid., 16.

82. My copy entitled "Assassination of Carlo Tresca" came from the George Arents Research Library, Syracuse University.

83. Salvemini, 17.

84. Ibid., 33.

85. Ibid., 203.

86. Memorandum in D.A.'s files.

87. Salvemini, 94.

88. Block, 196.

89. Alan Block (1980) East Side--West Side: Organizing Crime in New York, 1930-1950 (Cardiff, Wales), 197.

90. Ibid., 198.

91. The information on Marinelli is in Thomas E. Dewey, <u>Personal Papers</u>, Series 1: Early Career, Box 90, The University of Rochester Library, Department of Rare Books.

92. Block, 73.

93. Cannistraro's Introduction in Salvemini, xxx.

94. Salvemini, xxxii.

95. Memorandum in D.A.'s files.

96. Gabriel Kolko (1968) <u>The Politics of War: The War and United States Foreign Policy, 1943-195</u> (New York), 20.

97. Kolko, 21.

98. Kolko, 22.

99. Kolko, 42.

100. Kolko, 43.

101. Ibid.

102. Kolko, 57.

103. Ibid.

104. Kolko, 60.

105. See Rodney Campbell (1977) <u>The Luciano Project: The Secret Wartime Collaboration of the Mafia and the U.S. Navy</u> (New York).

106. History Project, STrategic Services Unit, Office of the Assistant Secretary of War (1976) <u>The Overseas Targets: War Report of the OSS (Office of Strategic Services)</u>, Vol. 2 (New York), 63.

107. History Project, 81.

108. Block, 109-113.

109. See S.J. Woolf (ed.) (1972) <u>The Rebirth of Italy, 1943-50</u> (London); Ronald Radosh (1972) <u>American Labor and United States Foreign Policy</u> (New York); and Fred Hirsch and Richard Fletcher (1977) <u>CIA and the Labour Movement</u> (London).

110. Among the latest publications consult Peter Wyden (1979) <u>Bay of Pigs: The Untold Story</u> (New York).

Chapter 8
Violence, Corruption and Clientelism
The Assassination of Jesus de Galindez, 1956

Published in Social Justice (1989).

My subject is the political murder of Jesus de Galindez Suarez, mildly characterized by a friend as "a Basque scholar who was studying and lecturing at Columbia University in the early and middle 1950s." Galindez who had spent several years in the Dominican Republic during World War II was finishing his doctoral dissertation on the first 25 years of the administration of Rafael Leonidas Trujillo y Molina, dictator of the Dominican Republic from 1930 to 1961. Around March 12, 1956, Galindez was indubitably murdered, although the exact time and location of his death is uncertain.(1) He was last seen on that late winter day by a student who drove him from the University to a subway station in Manhattan. Galindez vanished from New York, spirited to the Dominican Republic. But whether he made that journey dead or alive is not precisely known.

It is not my intention to produce yet another indictment of Trujillo's monstrosity, about which so much is already known, although it is inevitable that many of his crimes must be enumerated. I wish, instead, to discuss the hidden core of U.S./Dominican relations which are revealed by a careful study of the Galindez assassination. These relations were grounded in an admixture of secret services, political corruption, and political violence that bound the two states together throughout Trujillo's regime and beyond. In fact, after Trujillo's own assassination engineered by the CIA in 1961, and after American occupation of the Dominican Republic in 1965, and after several years of renewed American tutelage the "Santo Domingo newspaper El Nacional last December 30 filled a page and a half of newsprint with the details of 186 political murders and thirty disappearances during 1970." The newspaper added that "The Dominican terror resembles the current wave of political killings in Guatemala . . . in that the paramilitary death squads are organized by the armed forces and police, which in both cases over the years have been given heavy U.S. material and advisory support.(2)

VIOLENCE AND THE NATION-STATE

My theoretical text is primarily derived from Anthony Giddens' The Nation-State and Violence which pivots to a substantial degree on a proper understanding of "surveillance." He writes, for example, that "Information storage is central to the role of 'authoritative resources' in the structuring of social systems spanning larger ranges of space and time than tribal cultures," and that surveillance is "the key to the expansion of such resources." The term surveillance includes the "control of information and the superintendence of the activities of some groups by others."(3) Giddens deliberates on the distinction between totalitarian rule, which immediately comes to mind when speaking of such matters, and a "'normative theory of political violence'."(4) Totalitarianism, he sees as a "tendential property of the modern state," despite its almost universal application to movements, parties, leaders and ideas expressed in Fascist Italy, Nazi Germany, the Soviet Union under Stalin, and more recently the Pol Pot regime in Kampuchea. Totalitarianism, indeed terror, rests upon the effectiveness of surveillance which, Giddens reckons, "must be regarded as an independent source of power, maximized in the modern state, ... And as surveillance is ubiquitous, as "aspects of totalitarian rule are a threat in all modern states, even if not all are threatened equally or in exactly the same ways," then it follows that totalitarianism is a "tendential property of the modern state."(5) Hannah Arendt had argued that terror was the basis for the rule of totalitarianism;(6) Giddens, instead, cleaves to the sociology of surveillance as "primus inter pares"--the control of information produces mass support which "generates the political leverage within which terror can be used against categories of 'deviants'."(7)

The reformation of surveillance is part of the "modernity" process in which the "industrialization of war" is also vital. The term covers those related changes stemming from the industrial production of arms, along with the development for military purposes of innovative methods of transportation and communication. Technological transformations of such magnitude wrought significant modifications in the sociology of command and military service: officer corps and armies were reorganized and professionalized.(8) Giddens argues that these linked occurrences shaped the nation-state system and led "to the creation of a world military order that substantially cross-cuts the divisions between 'First', 'Second' and 'Third' worlds."(9) This structure is propelled forward by "super-power hegemony," plus the global systems of alliance which include the training of military cadres by one or the other super-power.(10) It is this evolution that makes grotesque states such as the Dominican Republic more than simple tyrannies. They are that, of course; but they are also much more.

The American Discourse on Secret Police

Surveillance, political policing, totalitarianism, etc., were common terms in the U.S. no later than the end of the nineteenth century. Then they both described and helped shape events in the many battles between industry and labor as it sought to organize. The organization of labor was fought by secret political police in various guises in order to stop Communist subversion, the argument went, which would bring totalitarian government in its wake.(11) This characterization took on added gusto by the end of World War II when subversion by Communists was seen to be everywhere. The U.S. and the rest of the non-Soviet world was threatened by the insidious actions of secret agents as much as by an insidious treacherous ideology.(12)

The American political idiom, especially when it came to secret police, was framed by the shape it gave to the Communist enemy both without and within. And for a very long time the U.S. was able to maintain the framework, control the discussion, monopolize the moral marketplace, with revelations of Soviet Cold War perfidy, focusing on the brutality of the Communist secret services. This only became problematic years later when CIA secrets, its "dirty work," were uncovered.(13) Nonetheless, these revelations were handled in such a circumscribed manner that the U.S. was still able to hold, albeit shakily, the high ground. Two examples will make the point clear. The first is drawn from the mid 1960s when Americans still unblushingly pointed to Soviet Cold War crimes. The second occurs a decade later when the Senate investigated CIA crimes.

In the early spring of 1965, the Internal Security Subcommittee of the Senate Committee on the Judiciary focused attention on the serious topic "Murder and Kidnaping As An Instrument of Soviet Policy."(14) The hearing record was partially structured around information gathered from the West German trial of Bogdan Stashynsky held in October 1962. Stashynsky was a Ukrainian who had been recruited by the Soviet Ministry of State Security (MGB) as an agent. Initially, Stashynsky's task was to penetrate and provide information about a dissident group called the Organization of Ukrainian Nationalists. In 1954, the MBG became the Committee for State Security (KGB) and it ordered Stashynsky to Munich, West Germany, where he was to continue his spy work. From spying Stashynsky was next ordered to assassinate a Ukrainian living in Germany which he did in 1957. Two years later he murdered another. Stashynsky surrendered to West German and American authorities in 1961, apparently out of remorse.(15)

The Stashynsky trial, Senator Thomas J. Dodd stated in the Subcommittee report, "established for the first time in a court of law, that the Soviets employ murder as an instrument of international policy . . . and that political murder has now, as it were, become an institution." Even without the Stashynsky case, Dodd added, "every serious student of Communism" was fully aware that the Kremlin "maintains an international murder apparatus."

Dodd mentioned several murders from the 1930s and early 1940s in Switzerland, Paris, New York, Mexico City, and Washington D.C., and noted that the "free world had no doubt" these killings were carried out by agents of the Kremlin. To more firmly establish his point, Dodd added to his remarks about Communism and murder a statement made by the German Court during the Stashynsky trial. The Court said it was now "obliged to ascertain with regret that the political leadership of the Soviet Union . . . officially order and has murders carried out on German territory."(16)

If political murder was not bad enough, the Senate Subcommittee established that professional criminals were recruited by the Soviet secret services to carry out kidnapping and assassination. Details of this collaboration were provided by Peter S. Deriabin, a Soviet defector who had a long career in military counterintelligence with SMERSH, Okhrana, and finally in the MVD working in post-war Vienna. In addition to Deriabin's testimony on this point, other evidence was provided by a group known as the Committee To Combat Soviet Kidnappings" whose honorary chairman in the mid-1950s was the former U.S. Ambassador to the U.S.S.R., Admiral William H. Standley.

An example of the collaboration between professional criminals and counterintelligence agencies was reportedly the kidnapping of Dr. Walter Linse in the summer of 1952. Dr. Linse, who was residing in the American sector of Berlin was reportedly "then acting chief of one of the most effective anti-Communist organizations in West Berlin, the Association of Free German Jurists."(17) In its presentation of this case, the Senate Subcommittee included the following statement of Johannes Stumm, Police President of West Berlin:

> A thorough and widespread investigation
> . . . has resulted in the identification of four East
> Berlin professional criminals who stand accused
> of assaulting and abducting Dr. Linse. . . . The
> investigation has disclosed not only the names of
> the four principal kidnapers but also of 13 other
> hand-picked and professional outlaws and
> gangsters who played important roles in one of
> the most brazen and repugnant crimes in the
> history of Berlin. . . . These four men are
> identified as part of a criminally organized and
> criminally subsidized ring of kidnapers approved,
> sponsored, and directed by the GDR Ministry for
> State Security which has become widely known as
> the dread MMS, which not only is modeled after
> the MGB, the Ministry of State Security of the
> Soviet Union, but is an integral thriving organ of
> the Russian Police State. . . By the very nature

> of its criminal mission the MSS is compelled to
> rely completely upon murderers, dope addicts,
> highly trained burglars and black market
> operators. . . . The West Berlin police have
> unimpeachable evidence that this MSS-sponsored
> and protected kidnap organization is financed by
> the sale of great quantities of cigarettes, coffee,
> and silk stockings on the black market.(18)

The message of the Senate Internal Security Subcommittee's hearing published under the very provocative title **MURDER INTERNATIONAL, INC.,** (a play on the name Murder, Inc. which was the title given to a band of New York gangsters by a newspaper reporter in 1940) was unmistakable: the Soviet Union and its various surrogates employed murder as a political weapon against dissidents residing in foreign lands, and their Secret Services recruited professional criminals, especially it appears in Germany, to carry out many of these attacks. Moreover, political murder and the collaboration between agents and criminals emerged not just out of the chaos of war and reconstruction, but, it was claimed, had something fundamental to do with the nature of Communism.

Ten years later another Senate Committee, this time The Select Committee to Study Governmental Operations with Respect to Intelligence Activities, issued a report dealing with the same topic of political murder and assassination although the targets of these plots were not dissident emigres but Presidents and Prime Ministers. In 1975 the secret service under scrutiny was the Central Intelligence Agency not the KGB or MSS. The report of what was known as the Church Committee begins with a prologue which insists that the events investigated "must be viewed in the context of United State policy and actions designed to counter the threat of spreading Communism."(19)

Starting at the end of World War II, a new struggle emerged between what the prologue calls the "Free World" and the "Sino-Soviet bloc" which continually threatened "encroachment" upon the former. The policy designed to combat this menace was "containment" which meant among other things a "relentless cold war against Communist expansion wherever it appeared in the 'back alleys of the world'."(20) Tit for tat, "this called for a full range of covert activities in response to the operations of Communist clandestine services." Containment in the world's back alleys heated up when Fidel Castro triumphed in Cuba. This galling situation--"the first significant penetration by the Communists into the Western Hemisphere"--deeply affected American clandestine activities. In sum the prologue maintains,

> Throughout this period, the United States
> felt impelled to respond to threats which were, or
> seemed to be, skirmishes in a global Cold War

against Communism. Castro's Cuba raised the
specter of a Soviet outpost at America's doorstep.
Events in the Dominican Republic appeared to
offer an additional opportunity for the Russians
and their allies. The Congo, freed from Belgian
rule, occupied the strategic center of the African
continent, and the prospect of Communist
penetration there was viewed as a threat to
American interests in emerging African nations.
There was great concern that a Communist
takeover in Indochina would have a 'domino
effect' throughout Asia. Even the election in 1970
of a Marxist president in Chile was seen by some
as a threat similar to that of Castro's takeover in
Cuba.(21)

The Church Committee looked at the attempted assassinations of
Patrice Lumumba in the Congo (now Zaire), Fidel Castro in Cuba, Rafael
Trujillo in the Dominican Republic, Ngo Dinh Diem in South Vietnam, and
General Rene Schneider in Chile. And appallingly enough, in the
investigation of CIA plots against Castro, it was revealed that the Soviets and
East Germans were not the only ones to use professional criminals in covert
operations.

There were at least eight assassination plots aimed at Castro from
1960 to 1965, only one of which, the Senate incorrectly reported, involved the
"use of underworld figures."(22) In the summer of 1960, high officials of the
CIA (Deputy Director for Plans, Richard Bissell, and Director of Security,
Sheffield Edwards) used as their intermediary to the underworld Robert A.
Maheu an ex-FBI agent and burgeoning private security tycoon. Maheu, who
had already performed, as the Committee put it, several "sensitive covert
operations" for the CIA, contacted John Rosselli, a well-known organized
crime figure.(23)

For most of his criminal career Rosselli worked in California selling
bootleg liquor during Prohibition and later working the mob's racing wire
business in Los Angeles. He left his mark on the movie industry through an
extortion conspiracy against certain studio executives. He was convicted in
Federal Court for this enterprise just a day or two before Christmas, 1943.
Paroled in 1947, he expanded his activities to Las Vegas, in particular the
Tropicana Hotel and Casino. The FBI regarded Rosselli as a mob gambler
and general factotum for major racketeer Sam Giancana from Chicago.(24)

Following Rosselli's recruitment, the CIA brought two other
professional criminals into the assassination scheme. One was Giancana, the
other Santo Trafficante a crime boss from Tampa, Florida, with international

contacts. The first try at murdering Castro by this team failed, however.
After this, and the debacle of the Bay of Pigs operation, the Castro
assassination project was taken from Bissell and Edwards, and turned over to
CIA officer, William Harvey. He was supposedly an assassination expert who
only several months before had been "given responsibility for establishing a
general capability within the CIA for disabling foreign leaders, including
assassination as a 'last resort.'"   Harvey's venture was codenamed
EXECUTIVE ACTION and was later subsumed under the cryptonym
ZR/RIFLE.(25) This project's stated purposes were "to spot, develop, and
use agent assets for Division D operations." These, which likely originated
with the National Security Agency, were passed to the CIA, and called for
the securing of code and cipher materials.(26)

ZR/RIFLE was directed "against third-country installations and
personnel." The principal agent in charge of ZR/RIFLE field operations was
known as QJ/WIN, supposedly a "foreign citizen with a criminal background
who had been recruited by the CIA for certain sensitive programs." The
Church Committee mentions QJ/WIN because the CIA and this agent were
implicated in planning the murder of Patrice Lumumba. Luckily for the
Agency, Lumumba's political enemies in the Congo supposedly got to him
first and thus spared the CIA from the ignominy of real murder. Whoever
else may have been done in by Executive Action-ZR/RIFLE, when it came
to killing Castro, Harvey fared no better than his predecessors.

So dismally ended, the Church Committee mistakenly reported, the
collaboration between the CIA and professional criminals in the Castro
assassination episodes. The Senate panel was acutely distressed with the
CIA's "attempted" murders, and chagrined with its use of underworld figures
by agencies of the United States. Trying to deal with the latter issue, the
Select Committee gave several reasons why professional criminals must not
be used for "their criminal talents." These assignments gave mobsters an
upper hand with the government, granting them an ability to "blackmail"
when necessary to avoid prosecution. It also made the government a
potential accomplice in crimes committed solely to further racketeers'
interests. But most of all, the entire affair (assassination planning, secret
relations between the CIA and the underworld) was fundamentally improper-
-it undermined democracy.(27)

Separated by a turbulent decade, the two Senate reports on political
murders invite a brief comparison. They both note assassinations have been
a Soviet and American covert policy option since the Soviets started the Cold
War. But that doesn't quite mean an equality of shabbiness. As the 1975
report on the CIA points out, the U.S. secret service didn't actually kill
anyone. It tried but failed or was just beaten to the punch by others, thereby
preserving, through ineptitude or the "good grace of God," a vital distinction
between America and Russia. The Soviets and their surrogates in places
such as East Germany really killed as the 1965 Internal Security

Subcommittee reported. It is a tendentious point no matter how argued, lacking the elementary sense about political violence enunciated by Trotsky (a notable victim of violence himself) decades earlier. He wrote that "history down to now has not thought of any other way of carrying mankind forward than that of setting up always the revolutionary violence of the progressive class against the conservative violence of the outworn classes."(28) With this formulation at least one can choose violent sides.

CLIENTELISM

The Dominican Republic has been and remains a client state of the U.S. Typically used to describe a form of political behavior most noticeable in "patron-client relationships in small rural villages" in which reciprocity and the inequality of status produce an asymmetrical exchange relation, clientelism now characterizes ever larger structures of inequality.(29) Clientelism is at the root of the Mafia phenomena in Sicily and Calabria, and in the international relations between super-powers and third world states linked together in the world military order.   In the asymmetrical exchange relations characteristic at this level, the super-power trains and arms the local military whose job is domestic stability meaning the rooting out of all left-wing subversives. This is done to shore-up traditional forms of economic exploitation which represent another vast ocean of violence and inequality linking foreign economic interests and local notables in still other asymmetrical relations.

The history of Hispaniola, the island shared by the Dominican Republic with Haiti, is one of relentless misery--"ill fortune has been the lot of the islanders ever since the day Columbus disembarked from the Santa Maria," one low-keyed historian remarked.(30) It was a slave island torn between the rival imperial ambitions of France, Spain, and other European colonial nations, always subject to the ravages of buccaneers. The French dominated the western third of the island which was known as Saint Domingue, the Spanish the rest. This lasted until the 1790s when French dominion momentarily triumphed over all of Hispaniola. From that decade on through part of the next, however, the most famous slave revolt in the Americas, under the leadership of Toussaint L'Ouverture, Jean Jacques Dessalines, and Henri Christophe, established Haiti (the western third) as a free nation. Free but profoundly poor, Haiti sank deeper and deeper into economic squalor. This wretchedness was assisted by its former master France which demanded huge cash reparations for the dispossessed planters and a sort of favored-nation trading status before it finally granted Haiti diplomatic recognition in 1825.(31) Meanwhile, the larger eastern Spanish part of the island fared only marginally better. It too was clawed at by imperial nations, and its newly freed neighbor during much of the nineteenth century. In fact, Haiti united and ruled the island from 1822 until 1844, when the Dominicans finally secured their liberty.

Even this measure of control didn't last without a lapse here and there. Some Dominicans were dissatisfied fearing that without national security political freedom was valueless. They were convinced it was desirable to exchange national sovereignty for security under some foreign flag. The Dominican Republic shopped for a buyer. Spain finally agreed, and reappeared in 1861 as master of the nation. Spanish colonial troops were shifted from Cuba and Puerto Rico for emphasis. This experiment was short-lived, many of the soldiers died of yellow fever, and by 1865 Queen Isabella II called it off. The desire for foreign rule as the only method of internal stability was not silenced with the Spanish fiasco, however. "Once rid of the Spaniards in 1865, the witless rulers of the republic blundered into various schemes for surrendering all or part of the republic's sovereignty to the United States." Senator Charles Sumner, the radical republican from Massachusetts, warded off the plan and its backers.(32) These island struggles left the unfortunate legacy of unrelieved enmity and violence.(33)

Hispaniola was like a cursed land. Whether Haitian or Dominican, rulers were invariably tyrants, maniacs, incompetents, who, with few exceptions, exploited exorbitant and searing racial distinctions and class divisions. Standing over this potpourri of affliction was foreign capital. Before the Haitian revolution, the French portion had an economy dominated by king sugar.(34) Colonial sugar production started in Barbados--"the brightest jewel in the crown for the English long before India took over that description"--then moved to Jamaica and finally Saint Domingue which had about half a million slaves by the 1780s.(35) Historian Eric Williams described it as "the worst hell on earth in 1789, . . . absorbing 40,000 slaves a years." As expected, the triumphant slave revolution totally decimated the sugar industry thus unavoidably furthering Haiti's plummet into destitution. Saint Domingue's place in sugar production was taken by Cuba and soon greatly enlarged.(36)

Until the 1870s, sugar played a very minor role in Dominican life. During that decade its cultivation on a fairly large scale was introduced by expatriate Cubans. But it was American capital, especially in the period between 1897 and 1930, which transformed Caribbean sugar businesses, including those in the Dominican Republic, into a modern industry. Although Cuba and Puerto Rico were the stars of the American industrialization of sugar, the Dominican Republic's reduced scale experience was also consequential. Two American companies controlled the Dominican industry: the South Porto Rico Company had 150,000 Dominican acres, and the West Indies Sugar Corporation another 100,000. Moreover, they owned six of the 14 Dominican sugar mills which because of the marked differences in mill output meant control of three/quarters of Dominican production by 1939. The industrialization of sugar in the Caribbean was keyed to the vertical integration of the industry--the sugar companies had huge cane growing factories, and their own modern railroads for moving the product to their own warehouses, wharves and steamers. Other ancillary industries cropped up here and there to service the industry. In the Dominican Republic, the American sugar juggernaut represented 67 percent of Dominican exports in the decade 1929-1938. In 35 years its output had increased by a factor of ten.(37) Sugar also enabled the Dominicans to exact a large measure of revenge against their perennial enemy Haiti. Haitian labor was dirt cheap and the Dominican Republic like Cuba and other Caribbean states imported Haitians to work in the cane fields. They were paid dog wages and always cruelly treated.

Dominican clientelism is only partially covered by noting U.S. hegemony over its most important industry. In an even more direct way than so far noted, Americans grasped parts of the economy until all of the Dominican Republic was truly theirs. This larger development began during the regime of Ulises Heureaux (1882-99) when American entrepreneurs organized the San Domingo Development Corporation. They extended some credits to Heureaux and in return were given control of the nation's customhouses. This cushy arrangement suffered when the dictator was assassinated in 1899, and chaos returned. Soon more than $32 million was owed to foreign nationals, and European nations threatened occupation or at least control of the customs revenue. To forestall this, and to protect the inflated claims of the San Domingo Development Corporation, the U.S. secured control of the most important Dominican customhouse in 1904. The Dominican situation prompted President Theodore Roosevelt to issue his famous "Corollary" to the Monroe Doctrine which held the U.S. would reluctantly police the hemisphere acting in cases of flagrant "wrongdoing or impotence." The following year the U.S. seized all Dominican customhouses and took over all revenue collection. Fifty-five percent was given to satisfy foreign creditors, the rest was remanded to the Dominican Treasury.(38)

The customs receivership was the first dramatic instance of U.S. intervention. Eleven years later, after more endemic internal discord, U.S. Marines who were stationed in the country were ordered by President Woodrow Wilson to take control. The Dominican Republic was placed under military occupation, government and law. The occupation which lasted for eight years, joined with the industrialization of sugar in conveying a radically skewed modernity. American officials created roads, initiated public works projects, and drastically improved sanitation, communications and education.(39) Ruling by decree, Americans also organized the public treasury, and tried to streamline the government bureaucracies by firing "useless employees." These achievements, essential for the movement of goods and troops, were set alongside the most important advancement of all. Americans trained and armed a national constabulary.(40) Within a short span, the Dominican Republic had become an American client state and part of the modern world military order as outlined by Giddens above. A modern military and industry had been created which were forever indebted to (and in debt to) the U.S. These had been crafted on a third world nation--even in 1968 it was reckoned that "the majority of Dominicans are without adequate food, water, and housing, have no medical or health facilities, no educational or recreational opportunities, no electricity, insufficient land, and above all, no hope"--with an enormous penchant for violence, and a traditional political structure composed of a "hierarchy of despotisms."(41)

TRUJILLO

The career of Rafael Leonidas Trujillo y Molina was a product of the Marine Corps occupation. He was born in 1891 in the little town of San Cristobal about eighteen miles west and slightly inland from the capital, Santo Domingo. The third child of eleven in a family characterized as upper middle class but only "within a small and isolated rural community, no element of which could aspire to high rank in the country at large," Trujillo's early upbringing appeared normal except perhaps for an excessive passion for neatness and cleanliness.(42) At the age of sixteen he took his first real job as a telegraph operator acquired through the influence of a relative. He was, however, more interested in petty criminality than anything else. Around 1916 he joined a gang known later as "The 44" which robbed small "bodegas," and engaged in minor graft and extortion. While running with the "44," Trujillo also worked for a couple of years in the sugar industry. He began weighing cane but it was antithetical to his style; he changed to a "guarda campestre," a kind of private policeman.(43)

In 1918, he finally found his true calling, one toward which he had been gravitating for some time. Trujillo enrolled in the National Police that December receiving his commission as a Second Lieutenant on January 11, 1919. Equivalent to a National Guard, the Police was directed naturally by the U.S. Marine Corps. Among their duties was suppression of rural bandits, or as some styled them peasant guerillas fighting the Marines, disrupting the sugar estates. Called "Gavilleros," they were eventually overcome.(44) Trujillo had an instinct for both counterinsurgency and clientelism--he cultivated the patronage of Marine Colonel Thomas Watson an important occupation official--and rapidly moved up the police/military ladder. He also formed an abiding association with Marine officer Charles McLaughlin who became an intimate companion and counselor. By the time the Marines pulled out in 1924, without McLaughlin who remained, Trujillo was Chief of Staff of the National Police which in the spring of 1928 officially turned into the National Army.

With his military power established and unquestioned, the Army his "private instrument of coercion and terror," Trujillo began actively scheming for the Presidency.(45) His campaign was both subtle and brutal. Winning, of course, was paramount but so too was U.S. recognition of his government. Through a series of stratagems he was assured of U.S. support. Greatly relieved he plunged forward having the Army and a gang of terrorists called "The 42" (whose leadership reached back to the youthful "44," and whose organization and personnel overlapped with that of the National Police) assassinate political opponents. On August 16, 1930, Trujillo was elected president. The prerequisite for the hierarchy of despotisms was in place; now it would express and refine itself in one cruel fashion or another over several decades.

One essence of the hierarchy was the elevation and enrichment of the family Trujillo which, in time, included Alma McLaughlin, daughter of Trujillo's Marine confidant. She married Trujillo's brother Hector, in 1960 after a fourteen-year engagement and the birth of a son. The marriage had been long delayed by Trujillo's wife, sensitive about sharing the Dominican spotlight with her.   Hector was his brother's pawn used in many ways including "standing in" for Trujillo as the President of the Dominican Republic in 1952. The puppet Hector also won re-election in 1957 when he ran as the only candidate. Hector was, of course, amply rewarded supposedly accumulating a fortune estimated at around $150 million.(46)

Trujillo himself owned, usually through "false fronts," the cotton and rope industry, the country's only oil company, major shipyard, milk processing plant, brewery, airport, airline, and Dominican banks, hotels, power plants, the major insurance firm, the newspaper "El Caribe," and so on. He also invested money abroad especially in U.S. companies in construction, oil, transportation, finance, insurance, hotels, sugar, and real estate in Manhattan, New Orleans and Miami. He was rumored to have extensive land holdings in Cuba, mining interests in Canada, and a fortune stashed in Swiss banks.

His jealous wife, Maria Martinez de Trujillo, had a controlling interest in Atlas Motors (Dominican importers of General Motors cars), a medical supply outfit, and was a sort of government loanshark. She was reportedly in charge of an organization which lent money at exceedingly high rates to government employees strongly encouraged to borrow. One of Trujillo's sons, Rafael Leonidas commonly called Ramfis, partially owned the Dominican television station (probably in partnership at some point with his uncle Jose Arismendy called Petan), a large cement factory, a peanut oil processing plant, a lumber plant, and several large farms.

Petan also had extensive real estate interests in the Dominican town of Bonao and cattle ranches in the Cibao region where he maintained order through a private militia. Another brother, General Pedro V. Trujillo Molina, who had a reputation for stupidity and drunkenness, owned a slaughterhouse controlling meat exports to Miami and San Juan.   Still another brother (it was after all a family of eleven children) known as Pipi ran the prostitution racket in the capitol, while a sister concentrated on urban real estate (over 500 houses) and cattle ranches. Her houses were rented to working class people; the rent collected by a "goon squad." A Trujillo nephew, Romeo Trujillo Lora, was a partner in a construction company which built schools and hospitals under government contract.

Another nephew became a medical doctor without ever attending school. Accompanied by armed friends, he took his medical exam from the thoroughly intimidated Medical Examination Board which merely asked him to identify "an eye, a nose, and an ear."

In order to usurp the economy, though this was hardly the only motivation, Trujillo created a terror state in traditional terms and a "surveillance state" in Giddens' terms: managing information and superintending the activities of various groups by the use of numerous clandestine ones. Trujillo developed several "cliques" called by one writer shifting bodies of henchmen chosen for their personal loyalty but kept atomized and isolated to prevent their garnering too much power.(47) These methods created a familiar form of stable instability. Because it was never entirely clear who had Trujillo's real confidence, his cronies' endemic insecurity drove them to work harder and harder to please, to prove their loyalty. Under the Trujillo regime, citizens of the Dominican Republic naturally were quite a bit more insecure, exhibiting the sudden disturbing silences and clammy apprehensions typical of a terrorized populace.(48) They had good reason to be fearful as the regime didn't hesitate to assassinate political enemies at home although not as many were killed as one might think. A few exemplary killings appear to have been sufficient to make the point. This wasn't the case with Haitians, however. In 1937, Trujillo's forces slaughtered around 18,000 along the Haitian/Dominican border.(49)

The surveillance of Dominican society went far beyond the nation's borders to include a watch on Dominicans living abroad. Additionally, Trujillo's spies infiltrated other Caribbean secret services, newspapers, radical and revolutionary groups. A great deal of this effort was undertaken in Cuba, Puerto Rico, Mexico, and Venezuela which broke diplomatic relations with the Dominican Republic in 1945 when a radical government led by Romulo Betancourt came to power. Betancourt despised Trujillo and actively plotted with others including, of course, exiled Dominicans to bring it down. Trujillo's reaction was to turn up the heat on Betancourt by planning unsuccessfully his assassination.

Venezuela's effort was aided by radicals in Cuba, another Caribbean hotbed of political intrigue touching on indigenous and regional issues. Among the many ways Trujillo dealt with Cuba in the 1940s was through infiltration and subversion of Cuban organizations particularly those in league with Venezuela. He sent his sub-secretary of Labor to Havana, for instance, to try and cut a deal with the Confederation of Workers of Cuba. The alleged purpose was to secure help from the Cuban organization in creating a national labor union in the Dominican Republic.(50) The real purpose was to create another outlet for infiltration also useful for "agents provocateurs." Additionally, Trujillo courted Cuban pilots during the 1940s; several had secret meetings in the Dominican Republic with Army General Fiallo.(51)

In contemporary international clientelism, it is imperative to have the most intimate knowledge possible of a client's intentions. Precipitous actions by a client state can be politically dangerous for its patron. Inscrutability used as a internal, domestic, coercive technique cannot be allowed to cross over and cloud the international realm. Therefore, Dominican plotting was carefully watched by the FBI which had extensive spy networks throughout the Caribbean, Central and South America. The FBI knew in March 1946, for example, that conspirators were receiving financial assistance from Trujillo in order to overthrow the Betancourt regime in Venezuela (FBI, "Correlation": 21). The next year the newly-created Central Intelligence Agency was on the scene adding to the U.S. surveillance capability. By the end of the Agency's first year, it reported on several Americans working for Trujillo as Aviation advisors and instructors.(52) The U.S. Army was not left out of the Caribbean spying trade, fast resembling a booming post-war "cottage" industry, and G-2 (Army Counterintelligence) was busy those same years keeping tabs on a number of Trujillo plans including one to conquer Haiti.

Obviously, there is nothing ephemeral about surveillance activities and organizations. They are now elemental, patrolling the inner space of the national territory whether in democracies or totalitarian states although clearly their methods differ, are more or less restrained, more or less covert, depending on the nature of the state. And, as we have seen, they patrol abroad penetrating and neutralizing threatening dissident groups, setting up false front propaganda organizations, countering the efforts of other secret services, and every once in a while assassinating opponents.

To say the very least, this is a very complicated business. Part of the complexity lies in the structure of secret police organizations which are rigidly compartmentalized, and usually in an internal state of near war. The feared ubiquitousness of double and triple agents, in addition to the structured secrecy, keeps intraorganizational groups in the secret services deeply suspicious, innately hostile, toward each other. Distrust within a service is further multiplied when foreign jobs in "friendly" nations are carried out because the operational groups typically utilize secret agents from the "friendly's" services who may have been doubled or are "agents of influence," etc. To further complicate these matters, there are usually third parties--invariably private detective firms many created by the political police for cover--used to provide a layer of insulation from public scrutiny if things go wrong.

Many of the relationships among secret political police organizations (both foreign and domestic) and private detectives have been built over time out of a variety of often boring routine needs. But there are times when these are replaced by others calling for savage and yet delicate missions. The political assassination of a foreign notable in a patron state such as the U.S. is the sort of job which requires precision, care, and a mix of foreign and domestic operatives. It is also important to be able to manipulate Congressional and public opinion. Despite all these difficulties and apprehensions, the unknown and dangerous contingencies which do exist and can upset the best laid plans, political assassinations in the U.S. are, of course, accomplished. In the 1950s, the Trujillo regime assassinated two particularly bothersome troublemakers living in New York. These murders were carried out by elements of his then loosely structured espionage services whose field commanders were a combination of high government official and old-fashioned thug, and Americans from the public and private clandestine services. The coverup following the second murder was financed by Trujillo and handled primarily by bent lawyers, corrupt politicians, and ambitious liberals most connected one way or another to the secret political police.

VICTIMS

The first killed was Andres Requena; the second Galindez. Lured to a building in lower Manhattan, Requena was shot to death on October 2, 1952. Born in the Dominican Republic, Requena rose to become Secretary of the Dominican Embassies in Madrid, Rome, Lima, Peru and Santiago, Chile. He left the Dominican diplomatic corps in 1944 and moved to the U.S. After a stint in the Army, he became an American citizen. Requena wrote for a somewhat moderate anti-Trujillo Spanish-language publication. But while restrained concerning Trujillo, it dealt harshly with Felix W. Bernardino, Dominican Consul General in New York in 1952, and, according to the FBI, his sister, Minerva who was the Dominican Minister to the United Nations.(53) Bernardino was the earliest suspect in the Requena killing.(54) Six weeks after the murder, the FBI was advised that Ramon Enrique Gallardo, a former officer of the Dominican army who travelled on a Dominican diplomatic passport, was the gunman acting under Bernardino's control.(55) That story fizzled out and was replaced years later by one which held that three American gunmen, one "a Detective Lieutenant of the New York City Police Department," had done the job for Bernardino.(56) A still later account states that Bernardino, who had a criminal record and was thought as early as 1950 to be a Dominican gunman, set up and shot Requena by himself.(57) The last version that I'm aware of held that American organized crime figures working for Bernandino's successor as New York Consul General were the killers.(58)

While many of the details about the Requena murder are unresolved, the same is not the case for the killing of Galindez. Galindez was born in Madrid on October 12, 1915, and graduated from the University of Madrid Law School. He served in the Republican government as the Legal Attache of the Basque Legation in Madrid, and then as an official in the Spanish Bureau of Prisons. During the Spanish Civil War he was an army officer. With the triumph of Fascism, he fled first to France and then in 1939 to the Dominican Republic. There he was employed as a legal adviser in the

Dominican labor department. Feeling harassed by the dictatorial regime, Galindez departed for the U.S. on the last day of March, 1946.

He still wasn't settled, however. He stayed in the U.S. for two years and then moved to France. About nine months later, according to one version of his movements compiled by the FBI, he returned to the U.S. and registered with the Justice Department as a "representative of the Basque Government-in-exile." In a short span of time he became the leader of the Basque Delegation to the United States. Galindez lived in the Delegation's quarters on Fifth Avenue in New York and worked as a part-time Lecturer in the Spanish Department at Columbia University. At some point he began to prepare for a PhD in political science. In the winter of 1956, at the age of 40, he presented his dissertation which critically judged Trujillo's rule over a 25-year period. This was the crux of the matter, the cause of his murder, his crime against Trujillo. Thirteen days later, March 12, 1956, Galindez vanished.(59)

The outcome of this murder was different in several ways than most of the Dominican regime's previous ones including Requena's. It produced an outpouring of American media and Congressional attention. The stories were the result of the disappearance itself, which created a sensational outcry from various left-wing Hispanic groups resident in New York and Washington, left-wing and liberal groups in general, and finally the mainstream press. Each pressured Congressional representatives in their different ways. Moreover, the Galindez murder also involved several Americans including a young pilot, Gerald Lester Murphy. He was suspected of flying the kidnapped and doomed Galindez to the Dominican Republic. The unlucky Murphy also disappeared, last seen in the Dominican Republic on December 3, 1956. Murphy's fate materially added to what was already a sensational story. These developments forced the Trujillo administration to put forth extraordinary measures to try and escape blame, to coverup the multiplying crimes. This venture revealed yet another layer of secret Dominican/U.S. relations, another stratum to the structure of clientelism.

The Nature of Galindez: Spy Versus Spy

The Galindez affair has still other secrets to tell. U.S. efforts to plumb the labyrinthine world of communism during and after the Second World War called for the recruitment of numerous well-placed "Confidential Informants," one of whom was Galindez. The fact that he was an informer was divulged by the FBI in a "secret memo" to the Attorney General and two Assistant A.G.'s, nine days after Galindez was kidnapped. It was then elaborated on when President Eisenhower requested "a fill in on [the] disappearance of De Galindez" a few weeks later. Ike may have been particularly anxious about this case because his wife's brother was currently in a business venture with Trujillo.

The Bureau noted that Galindez had initially provided information to its Legal Attache in the Dominican Republic during the war. His real service, however, started in September 1949. Within a few months he was "designated a security informant," given a codename, **NY 507-S**, and put on the Bureau's rather modest payroll--no more than $125 for services and $30 for expenses per month. Galindez furnished invaluable information (detailed and thorough written reports) on "Spanish and Latin-American matters in New York, including Spanish communist activities, Spanish Republican matters, [the] Nationalist Party of Puerto Rico, Dominican and other Latin-American revolutionary activities," the Veterans of the Abraham Lincoln Brigade, and the Joint Anti-Fascist Refugee Committee.(60)

Keeping nothing back from the FBI, Galindez told his control about his Trujillo dissertation early in 1955. He was warned to find another topic, that the FBI could not protect him from Trujillo's thugs, told to reflect upon the Requena murder, and finally put on notice that "his services might have to be discontinued if he became involved in a controversial situation with the Dominican Government." He, nonetheless, decided to push on with the dissertation believing that at most a campaign of slander might be launched against him by the Dominican regime and its many friends.(61)

That Galindez was a professional informant validates, if validation is still necessary, the pervasiveness of surveillance as one of the consequential features of the new world order. And in this world of "spy versus spy," the most difficult job is untangling the overlapping and intertwining networks of influence and exchange in order to clearly see the organization of surveillance, the construction of clientelism. To borrow a metaphor from studies of counterintelligence which indicate how difficult this task is, these networks are encased within a "wilderness of mirrors."

The timing of Galindez' final entry into the U.S. is a case in point. The FBI document prepared for briefing both the Attorney General and President Eisenhower, and used above to reconstruct Galindez' wanderings, has him in New York working as an informant in 1949. But another FBI document tells a slightly different story. The text notes that Galindez left the U.S. in March 1950 for both Venezuela and Havana on Basque Government business and returned sixteen days later. For unspecified reasons, Galindez' return was blocked. He was only admitted under a "certain recent clause under the immigration statutes, whereby the Central Intelligence Agency was granted the privilege to bring to the U.S. one hundred aliens a year, whenever such aliens have been or can be of assistance to the U.S." This may mean that Galindez served the CIA as well as the FBI.

In this regard, a fairly recent publication, unfortunately without citations, claims the CIA invested more than $1 million dollars in Galindez in order to finance his work as head of the Basque government-in-exile.(62) Part of this story was likely gathered from the work of Edward J. Mowery, a reporter for the New York <u>Herald Tribune</u> at the time of the Galindez assassination.   Mowery stated that Galindez represented the Basque movement and had collected millions in its name and then deliberately mishandled the funds. Subsequently though, Mowery was fired for his part in a clever Dominican "disinformation" campaign which claimed Galindez was alive and well in Eastern, "Communist" Europe. One of Mowery's prime sources for his Galindez/Basque material was Stanley Ross editor of <u>El Diario de Nueva York</u>, a well-known liberal Spanish language daily newspaper. But Ross, a Galindez friend and associate, indeed the individual responsible for reporting Galindez missing to the New York police, was also compromised by Trujillo. In fact, Ross may have been a deep-cover Trujillo agent for some time, part of the Dictator's extensive American spying apparatus. Ross, for his part, was expelled from the Inter-American Press Association in 1956 charged with working for Trujillo's interests.(63)

A CABAL OF ASSASSINS

The squad of Dominicans and Americans which finally stalked and kidnapped Galindez in 1956 slowly began to form in the spring of 1954. That was when the Trujillo government tried to borrow a couple of FBI agents "to act as companions and bodyguards for President Trujillo on a European trip of one month's duration beginning May 26."(64) A former FBI agent, then Director of the Miami Crime Commission, was approached by the Dominican Consul General Pedro Blandino who wished help in arranging a private meeting with J. Edgar Hoover. The Bureau turned him down, instead referring him to the State Department. But Blandino didn't trust various people in the State Department and pressed for at least the use of former FBI agents. The FBI remained obdurate and finally Blandino took its advice and agreed to meet with Robert F. Cartwright, Deputy Administrator, Bureau of Security and Consular Affairs, State Department. Hoover, by the way, was quite angry about even this much FBI involvement and scribbled his displeasure on the obligatory explanatory memo when it reached him. He wrote: "Absolutely wrong. It was none of our business & we should not have injected ourselves."(65)

Blandino and Cartwright met in Washington. Sensitive to Trujillo's security needs, the Deputy Administrator stated he knew just the man for the job. Cartwright set it up and the following day Blandino and John Joseph Frank met in the State Department.(66) A graduate from Georgetown University's Law School, Frank was a man with an intriguing clandestine background. He had been an FBI Agent from 1941 to 1949. That year he moved into the CIA. Frank's association with the Agency started in October 1949 and supposedly ended in March 1951, except for an admitted one-month stint during June 1953. Otherwise, he worked for the Office of Price Stabilization between CIA assignments, and then as a private attorney--investigator.(67)

As one might expect, however, Frank's CIA adventures carried on well after 1953. During the very period when he was courted by the Dominicans, Frank was in business with Robert Maheu, mentioned earlier as the CIA's intermediary with organized crime in the attempted assassination of Fidel Castro. Maheu's background was similar to Frank's--Georgetown Law School (although Maheu didn't graduate), the FBI where he specialized in counterintelligence, and then into a very suspicious semi-oblivion for a few years, only to emerge with his own private detective agency closely attached to the CIA. In the spring of 1954, Robert A. Maheu Associates was formed and was immediately under contract to the CIA.(68) Maheu's office was severely compartmentalized, as researcher and writer Jim Hougan notes, "with different employees handling different assignments on a need-to-know basis" providing "a magnificent framework for deniability."(69)

Frank was handled in just such a manner according to Maheu--allegedly hired for particular tasks in 1954 and 1955 when there was a need for some special surveillance or "leg work" for a case.(70) But there was more to it than that. Frank, Maheu, and another co-worker at Maheu Associates, Thomas A. Lavenia, formerly of the Secret Service, tried to "revamp" Trujillo's security system using electronic equipment from Research Products of Danbury, Connecticut, an outfit connected to Maheu. One other who got involved in this deal on the American side was Allan F. Hughes, a "former Counter Intelligence Corps employee," employed by Maheu. Frank also was given an airline credit card by Maheu Associates, as well as temporary quarters in New York at the National Republican Club where Maheu kept a suite of rooms.(71)

Maheu Associates was quite busy in 1954 working one of the most spectacular clandestine jobs of the decade--the framing of Aristotle Onassis. The frame was the consequence of an agreement between Saudi Arabia and Onassis which directly threatened the International Petroleum Cartel, and was aided by Onassis's brother-in-law Stavros Niarchos, also a Greek shipping tycoon.(72) Niarchos and Onassis were prickly competitors from the start, but when Niarchos faced financial ruin and possibly prison for violations of the Ship Sales Act of 1946, he quickly joined forces with the oil cartel and betrayed Onassis. One of Maheu's assignments was to place a wiretap on Onassis's New York offices, a chore apparently carried out by Frank.(73)

While Frank was engaged on many clandestine fronts, he had succeeded in also entering the Dominican Military Intelligence Service (SIM) run at that time by Lieutenant Colonel Arturo R. Espaillat a graduate of West Point.(74) Espaillat served as Dominican Military Attache in Washington, D.C. in 1950, and accompanied Trujillo and his new American bodyguard Frank to Spain and Italy in 1954. Frank and Espaillat shared a cabin on the return cross-Atlantic trip and became fast friends.(75) Two years later, shortly before the Galindez kidnapping, Espaillat was appointed

Dominican Consul General in New York.(76)   Also significant in Trujillo's
overseas intelligence and special operations details was Manuel de Moya
twice Dominican Ambassador to the U.S. and in 1956 Secretary of State
without Portfolio.(77)

Early in 1956, Frank, Espaillat, Gerald Murphy (the pilot), and
Horace William Schmahl, a naturalized American citizen born in Germany
who headed a private detective firm in New York, and Espaillat joined
forces. Trujillo's decision to kidnap Galindez went to de Moya, Espaillat and
likely Felix Bernardino.(78) Espaillat contacted Frank to set up Galindez for
the snatch, and arrange for the next part of the plan, a secret flight to the
Dominican Republic. Frank used Schmahl's office and organization to put
a watch on Galindez. In early March 1956, Frank, Espaillat and Bernardino
went to an airport in Linden, New Jersey, to speak with Murphy. Shortly
after, Murphy and a friend rented a Beechcraft and added an auxiliary gas
system. On March 11, Murphy flew the plane to a small airfield, Zahn's
airport at Amityville, Long Island, although he informed New Jersey airport
officials that he was heading for Miami. De Moya also flew that day, but he
went south to West Palm Beach on his way to Miami. The following day,
which was March 12, the date of the Galindez kidnapping, Frank left New
York for West Palm Beach.

At about 10 p.m. that night, it appears that Ralph Molins, a cousin of
Arturo Espaillat, and George Waldemar Mallen, a Dominican agent under
Espaillat's control, either mugged Galindez themselves and drove him to
Amityville or had it done by American mob friends.   Molins was an
American citizen born in Puerto Rico and raised in the Dominican
Republic.(79) He was a contact man for Espaillat with certain American
organized crime figures associated with the International Longshoremen's
Association and "Bayonne" Joe Zicarelli an important New Jersey mobster in
the arms business. In 1967, Life magazine claimed that Zicarelli in the 1950s
had sold over $1 million worth of weapons to his friend Rafael Trujillo, and
had helped in the murder of Requena and the kidnapping of Galindez.(80)

On the morning of March 13, Murphy with his unconscious passenger,
and one other conspirator there to make sure Galindez didn't wake up and
cause trouble, landed at Lantana Airport, West Palm Beach. While the
regular tanks were filled, the attendant noted "through an open doorway a
person lying on the cabin floor, with another seated in the rear." It was over
on the thirteenth. Three days later Murphy paid $4,412 in cash for a new
Dodge convertible. When he returned his plane to New Jersey, Murphy filed
a phony flight report--New York to Miami to New Orleans to Jacksonville to
Miami to St. Petersburg to Miami to Nassau to Miami to Ohio and finally
back to Linden, New Jersey. After his disappearance, Murphy's "personal
flight record and flight log books" were located. They showed clearly that he
flew from West Palm Beach to Monte Cristi airport in the Dominican
Republic.

Murphy's time was running out even as he went to work for the Dominican national airline (CDA) and bought another new car in the Dominican Republic. One signal that his protection was gone was the murder on October 25 of Colonel Salvador Cobian chief of the Dominican Security Service. Cobian was apparently a particularly important patron of Murphy's. With Cobian gone, Murphy made plans to leave the Dominican Republic, return to America and marry his fiance, Sally Claire. He resigned his airline job but recklessly proposed one last entrepreneurial venture which needed Trujillo's O.K. Murphy was heading for the Palace on December 3, but never arrived. That next morning, the Dominican police found his abandoned car near the sea in the suburbs of Ciudad Trujillo.

The killings associated with the Galindez affair were now almost over. The only outstanding issue left was to account for Murphy's disappearance. Octavio de la Maza, a Dominican from a "well-known and well-placed family," who also worked as a pilot for the Dominican airline, was selected as the final stooge. In mid-December, Dominican officials arrested de la Maza on the urging of the American Embassy which reported he and Murphy were enemies. A few weeks later, de la Maza was dead. The Dominican government stated he committed suicide in prison out of remorse for having killed Murphy. According to a suicide note, Murphy was murdered by de la Maza for making homosexual advances. Of course it was all nonsense.

* * * * * * * * * * * * * * * * * * *

NOTES

1. Jesus de Galindez, The Era of Trujillo: Dominican Dictator (University of Arizona Press, 1973; edited by Russell H. Fitzgibbons), xi.

2. Noam Chomsky and Edward S. Herman, The Political Economy of Human Rights, vol. I, The Washington Connection and Third World Fascism (South End Press, 1979), 244.

3. Anthony Giddens, A Contemporary Critique of Historical Materialism, vol. II, The Nation-State and Violence (University of California Press, 1987), 2.

4. Giddens, 295

5. Giddens, 296-97, 302, 310.

6. Hannah Arendt, The Origins of Totalitarianism (The World Publishing Co., 1958), 305-497.

7. Giddens, 305.

8. Giddens, 223.

9. Giddens, 5.   There is a varied literature on the military and "modernization" in which many authors, like Giddens, point out that "the hesitation with which we have approached the study of the primary functions of armies," means "little systematic thought has been give to the political sociology of armies and the roles that military institutions play in facilitating the processes of industrial and political development." This particular quote is taken from Lucian W. Pye, "Armies in the Process of Political Modernization," in John J. Johnson (ed.), The Role of the Military in Underdeveloped Countries, (Princeton University Press. 1962), 70.

10. Giddens, 252.

11. See Frank J. Donner, The Age of Surveillance: The Aims and Methods of America's Political Intelligence System (Alfred A. Knopf, 1980); Sidney L. Harring, Policing a Class Society: The Experience of America's Cities, 1865-1915 (Rutgers University Press, 1983); G. R. Jeffreys-Jones, American Espionage: From Secret Service to CIA (Free Press, 1977); and, Richard O. Boyer and Herbert M. Morais, Labor's Untold Story (The United Electrical, Radio and Machine Workers of America, 1955).

12. See David Caute, The Great Fear: The Anti-Communist Purge Under Truman and Eisenhower (Simon and Schuster, 1978).

13. "Dirty Work" is the title of a collection of articles on some of the CIA's activities abroad in a reader by Philip Agee and Louis Wolf, Dirty Work: The CIA in Western Europe (Lyle Stuart Inc., 1978).

14. U.S. Senate Subcommittee to Investigate the Administration of the Internal Security Act and Other Internal Security Laws of the Committee on the Judiciary, Murder International, Inc.: Murder and Kidnapping as an Instrument of Soviet Policy (Government Printing Office, 1965), II.

15. Senate Subcommittee, vii-xv.

16. Senate Subcommittee, v, xiv-xv.

17. Senate Subcommittee, 28.

18. Senate Subcommittee, 34-35.

19. U. S. Senate Select Committee to Study Governmental Operations with Respect to Intelligence Activities, Alleged Assassination Plots Involving Foreign Leaders: An Interim Report (Government Printing Office, 1975), xiii.

20. Senate Select Committee, xiii.

21. Senate Select Committee, xiii.

22. Senate Select Committee, 71.

23. Senate Select Committee, 75. Rosselli claimed to have been born in Chicago in the summer of 1904, but that is doubtful--The FBI examined the "signatures on Report of Birth, filed in 1936, and on Affidavit of Birth, . . . [finding] signatures to be forgeries." U.S. Department of Justice, Federal Bureau of Investigation, "JOHN ROSSELLI," Field Office File no. 92-113, Bureau File no. 92-3267, 3 October 1962, 1.

24. That, of course, meant Rosselli was "in contact" with Judith E. Campbell, the mistress of both Giancana and President John F. Kennedy.

25. Senate Select Committee, 83.

26. ZR/RIFLE and QJ/WIN material from the Assassination Archives and Research Center, Washington, D.C. Under the Freedom of Information Act, nine pages of William Harvey's notes on Project ZR/Rifle were released. The description is drawn from page 1.

27. Senate Select Committee, 259-60.

28. Leon Trotsky, Terrorism and Communism: A Reply to Karl Kautsky (University of Michigan Press, 1961), xxxiv.

29. Judith Chubb, Patronage, Power, and Power in Southern Italy: A Tale of Two Cities (Cambridge University Press, 1982), 3.

30. Hubert Herring, A History of Latin America From the Beginning to the Present (Alfred A. Knopf, 1965), 423.

31. Cambridge Encyclopedia of Latin America and the Caribbean (Cambridge University Press, 1985), 282.

32. Herring, 438.

33. Herring, 436-37.

34. Eric Williams, From Columbus to Castro: The History of the Caribbean, 1492-1969 (Vintage Books, 1984), 109-110.

35. Cambridge, 202-03.

36. Williams, 245, 361.

37. Williams, 423, 431, 439-40.

38. Herring, 440-41.

39. Howard J. Wiarda, <u>Dictatorship and Development: The Methods of Control in Trujillo's Republic</u> (University of Florida Press, 1968), 9.

40. Herring, 440.

41. Wiarda, 4-6.

42. Robert D. Crassweller, <u>Trujillo: The Life and Times of a Caribbean Dictator</u> (MacMillan. 1966), 28-30.

43. Crasweller, 34-36.

44. Crassweller, 45-46.
       One who considered the "Gavilleros" to have been guerillas and social bandits in the Hobsbawm sense was an FBI informant whose recollections were reported in U. S. Department of Justice, Federal Bureau of Investigation, **GALINDEZ PAPERS**, <u>Trujillo</u>, "Foreign Political Matters," File 109-485, 23 February 1961.
       This FBI material, as well as all subsequent FBI citations, are available because of the Freedom of Information Act and the rather incredible dogged work of researcher Alan Fitzgibbons. It was Fitzgibbons who successfully fought for the full release of the Bureau's **Galindez Papers.** He is still battling the CIA over disclosure of its files. I read the **Galindez Papers** in the FBI's main quarters (the J. Edgar Hoover building) in Washington. FBI clerks and researchers working for the FOI branch were, without exception, exceedingly helpful.

45. Crassweller, 69.

46. FBI, <u>Trujillo</u>, "Foreign Political Matters," <u>passim</u>.

47. Wiarda, 70.

48. Galindez, 121-22.

49. Crassweller, 150-57.

50. FBI, <u>Trujillo</u>, "Correlation Summary," 26 September 1974, 24.

51. Ibid., 26.

52. Ibid, 21, 28.

53. FBI, **GALINDEZ PAPERS**, <u>Requena</u>, File 2-1358, NY 109-45, 5-12.

54. FBI, <u>Requena</u>, "Interview with Dr. Nuno Goico," 12 February 1963; and "Memo from W. A. Branigan to A. H. Belmont," 14 October 1952.

55. FBI, <u>Requena</u>, "Communication by Assistant Attorney General Charles B. Murray," 28 November 1952.

56. FBI, <u>Requena</u>, document dated 13 June 1961.

57. FBI, <u>Trujillo</u>, "Correlation," 34-35; and <u>Requena</u>, "Interview."

58. New York State, Joint Legislative Committee on Crime, <u>Report for 1970</u> (Albany), 77-78.

59. FBI, **GALINDEZ PAPERS**, <u>Felix Hernandez Marquez</u>, File 97-3250, "Letter from John Edgar Hoover, Director, Federal Bureau of Investigation to Director, Central Intelligence Agency," 4 February 1957; and FBI, <u>Hernandez</u>, "Memo from A. H. Belmont to L. V. Boardman," 26 April 1956.

60. FBI, <u>Hernandez</u>, "Memo Belmont to Boardman," 2.

61. Ibid., 3.

62. Darrell Garwood, <u>Under Cover: Thirty-Five Years of CIA Deception</u> (Grove Press, 1985), 111.

63. FBI, <u>Hernandez</u>, [two documents entitled] "Stanley Ross, aka Stanley Rosenberg, Registration Act," 6 July 1956.

64. FBI, <u>Trujillo</u>, "Memo Pedro Blandino, Consul General Dominican Republic, Miami, from L. B. Nichols to Mr. Tolson," 17 May 1954.

65. FBI, <u>Trujillo</u>, "Memo Blandino."

66. FBI, **GALINDEZ PAPERS**, <u>John Joseph Frank</u>, File 97-3293, "Airtel to Director FBI," 13 February 1957.

67. FBI, <u>Frank</u>, "Memo from L. H. Martin to A. H. Belmont," 28 August 1956.

68. Jim Hougan, <u>Spooks: The Haunting of America--The Private Use of Secret Agents</u> (William Morrow, 1978), 265-73.

69. Ibid.

70. FBI, <u>Frank</u>, part 179, WFO [Washington Field Office] 97-881, section titled <u>Robert A. Maheu</u>, 40.

71. Ibid, 45.

72. FBI, **GALINDEZ PAPERS**, Horace William Schmahl, File 97-3424, "Interview Report, Ambrose Capparis, file 97-1435, 11 March 1958, placed in the report on Schmahl, 3 April 1958, 75.

73. Hougan, 281-90.

74. Wiarda, 53; Crassweller, 332; and FBI, **GALINDEZ PAPERS**, Ida Elisa Espaillat, File 105-108183, "Internal Security--Dominican Republic," 26 March 1962, 1-13.

75. FBI, Schmahl, "Summary of Galindez-Murphy Case, from E. Tomlin Bailey, Director, Office of Security to J. Edgar Hoover," 30 December 1957, 3.

76. FBI, Espaillat, 2.

77. FBI, Schmahl, "Summary," 2.

78. The following reconstruction is based on FBI, Schmahl, "Summary" unless otherwise noted.

79. FBI, **GALINDEZ PAPERS**, Rafael Molins, File 105-64225, WFO 97-980, "Interview with Molins," 23 October 1963

80. New York State, 277-78.

PART V:
DILEMMAS AND AMBIGUITIES
IN CRIME CONTROL

Chapter 9
Crisis in the Internal Revenue Service

This is a history of an unprecedented conflict fought between elements of the Internal Revenue Service during the 1970s. It documents the issues dividing the Service, discusses the major battles, and urges an interpretation of the conflict which concludes the victors were wrong in fact and duplicitous in manner. The theme which the triumphant established as the principal cause for the confrontation held that the vanquished deserved their fate because of their neglect of law and civil liberties. It seems to me, though, that this celebration of law and procedure was tactical having little to do with what put this turmoil into motion. Once in motion, however, it is the case that legitimate and important questions of law and procedure were raised, although usually to mask disingenuous actions.

Donald C. Alexander was appointed Commissioner of the Internal Revenue Service in the spring of 1973 by a very distracted President Richard M. Nixon.(1) Alexander served until the Carter administration replaced him in February 1977. Under Alexander's stewardship, the I.R.S. endured a long, complicated, bitter struggle brought about by conflicting visions of what and who the I.R.S. should investigate and prosecute. Alexander wanted the Service's renowned Intelligence Division (created on 1 July 1919) to confine its activities to civil matters, dropping criminal investigative work altogether.(2) However, it was the latter which provided the Intelligence Division with its primary mission since its creation.

In 1973 the I.R.S. had basically two structures which peaked at the Commissioner's office in Washington.(3) There was a Washington based structure called the National Office with a series of Assistant Commissioners in charge of various Divisions, and a national structure composed of Regions, Districts, and then local offices which also reported directly to the Commissioner. In Washington, the Intelligence Division rested under the guidance of the Assistant Commissioner for Compliance. In the hinterland, Intelligence was represented by Assistant Regional Commissioners, and then District Directors. There were Intelligence Division Special Agents (the only agents so designated in the I.R.S.) in local offices and they reported to District Directors of Intelligence, and so on up the bureaucratic ladder.(4) This was I.R.S. organization except for the Inspection Service which handled internal security matters and had a tighter command structure between its segment at the National Office and its Regional Officers, and the Chief

178

Counsel's Office. This last was really more a creature of the Service's parent organization, the Treasury Department, than of the I.R.S. itself. Outside of the Commissioner's Office, only Compliance had a normal structured role to play in forcing major changes concerning the Intelligence Division.

One stated reason for Commissioner Alexander's animus towards criminal work stemmed from his unhappiness with the Intelligence Division's practice of targeting special groups on the presumption they were willful tax evaders. Those which received special handling were organized crime figures and narcotics traffickers. Alexander publicly opposed singling out gangsters and drug smugglers as special targets from the start of his tenure.(5) Upon taking office on May 25, 1973, he was determined to turn the I.R.S. away from programs, projects, and operations fashioned to crack down on the above criminals. He was especially at odds with the two-year old Narcotics Traffickers Program (N.T.P.) which utilized a National Office Selection Committee to choose appropriate individuals for investigation, although somewhat paradoxically he did claim the Service's "war" on drug merchants was effective.(6) He found equally as distasteful I.R.S. involvlement in federal Strike Forces. These units were established in 1966 in the Department of Justice to combat organized crime. Personnel was drawn from the I.R.S., F.B.I., Federal Narcotics Bureau, Customs Service, etc.(7) According to Alexander, the I.R.S. was the "major contributor of investigative manpower" to Strike Forces which he stated had a strong record.(8) Nevertheless, his policy was simple and straight forward: I.R.S. forces should be directed at the general public.(9)

This stance opened a battle within the I.R.S. whose effects are still felt. The Commissioner eliminated, for instance, Intelligence Division investigations into financial crimes by Americans carried out in places such as The Bahamas, the Cayman Islands, and The Netherland Antilles which guaranteed depositor anonymity under stringent bank secrecy laws. These investigations involved undercover work aimed at "potential or "suspected" tax evaders and thus, Alexander maintained, were totally inappropriate to the reactive, civil tax enforcement Service he wanted.

Commissioner Alexander's distaste for off-shore investigations was clear even before he officially took office. As Commissoner-designate in early 1973, he had proposed a small but telling change in the tax return format itself. Alexander lobbied hard for the elimination from the tax form of the question which asked taxpayers whether or not they had a foreign bank account. This Foreign Bank Account Question had been placed on tax forms only a few years earlier to encourage compliance by taxpayers considering the illegal use of secret foreign bank accounts.(10) Alexander was opposed in this move by the Director of the Intelligence Division. This important early disagreement influenced Alexander's decision to attack. Persistence paid off and the Commissioner finally succeeded in removing the question for the

1975 tax year. Obviously significant, the question in a somewhat adumbrated form returned in 1976, just prior to Alexander's departure.

Interestingly enough, Alexander's decision to remove the I.R.S. from organized crime work was made while many others in law enforcement were becoming increasingly alarmed by organized crime's financial transgressions. Since 1970, this issue had become part of an ongoing Senate investigation into massive worldwide swindles carried out by organized crime using off-shore banks in foreign tax havens to negotiate stolen or counterfeit securities.(11) The Senate's Permanent Subcommittee on Investigations urged law enforcement coordination at the same time Alexander was planning I.R.S. disengagement.

The vendetta which took place was carried out within the political environment of the Watergate process which greatly aided him in dismantling the Intelligence Division. He successfully portrayed it as a Division out of control, involved in projects and operations which smacked of political spying and harassment, peopled by Special Agents devoted to illegal and unfair excesses in their zeal to catch crooks. Alexander was joined in this description by individuals who were targets of Intelligence Division investigations. They were quick to employ the Commissioner's analysis and rhetoric claiming they were merely victims of a mendacious political police force.

Because criminal investigative work dealing with organized, sophisticated, crime always involves informants and undercover operatives, and therefore has a covert phase, it wasn't difficult to cloud the already murky picture. It was indeed foggy given the recent past which was only then being painfully and hesitatingly revealed. The Watergate scandals proved to many quite reasonable people that the I.R.S. and other law enforcement agencies on the federal, state, and local levels, were in the main utterly outrageous enemies of the people. This was particularly so under the whip of the Nixon administrations which had convinced senior I.R.S. officials to cooperate in very nasty political work during the early 1970s. In response to the President's urging, the I.R.S. organized a clandestine intelligence unit (intelligence in this case refers to function not the Intelligence Division) within the Service called the Special Services Staff (S.S.S.). "Hidden away in the basement of I.R.S. headquarters in Washington, as David Wise wrote, the S.S.S. targeted individuals whose cultural expressions and political convictions were deemed unpatriotic or otherwise unacceptable.(12)

The Internal Revenue Service's cooperation with Nixon's abuses of power is a black mark against it. But it was untrue, as Alexander contended to the delight of a host of criminal entrepreneurs, that Intelligence Division operations were cut from the same repressive mold as the S.S.S. or had base political motives. In Alexander's attack upon the Intelligence Division there was a blurring of both issues and the timing of events in order to take the fullest advantage of the Watergate theme. Political policing included any Intelligence Division actions which appeared overzealous, unseemly, and/or illegal. Investigative procedures became the alleged hallmark of political policing. Thus reprehensible political policing no longer required a political base (political targets and ideological motivations) only a suspect procedure now "ipso facto" a political crime. Determined to either tame or terminate the Intelligence Division prior to his leaving office, Alexander and his staff turned every found procedural mistake (large or small, meaningful or irrelevant) into a "post hoc" justification for the policy. At times, this called for them to testify falsely under oath about their motivation and actions.

While this was plainly a Washington brawl, the most important fighting emerged over Intelligence Division operations in The Bahamas which started long before Alexander was picked to head the Service. When Alexander acted against the Intelligence Division, he usually did so with an accusing finger pointing at Bahamian operations known as Operation Tradewinds and its offspring the controversial Project Haven.

TRADEWINDS

Special Agent Richard E. Jaffe working in the I.R.S. Miami Office, constructed in the mid 1960s an undercover operation in The Bahamas.(13) It lasted in one form or another for a decade. Jaffe acted because The Bahamas had become a favored haven for American organized crime and tax evaders. Through a long process of criminalization The Bahamas had become less a nation, more a "racket"--a continuing criminal enterprise run by politicians paid to allow an assortment of modern buccaneers to roost. Their racket became a linchpin in what is generally called the "Underground Economy," an economic system designed to bypass national revenue agencies. This subterranean financial system is composed of the loot from thousands of criminal enterprises conducted around the world, plus unreported and untaxed millions if not billions in profit from legitimate businesses. The Bahamas is the laundromat for this money; its banks, protected by secrecy laws, wash the funds which are repatriated in the U.S. or dozens of other countries as loans and investments from veiled off-shore entities. Operation Tradewinds designed to catch the malefactors regardless of party or wealth, proved over and over again the utility and necessity for an Intelligence Division presence in The Bahamas.

In a relatively short time--1962 or so until the latter 1970s--The Bahamas developed into an international banking center with almost 350 banks and trust companies headquartered there.(14) By then, the little island nation was a principal Eurobanking hub, with Eurodollar assets totalling more than $100 billion. The Bahamas served as a locus for either booked or actual interbank exchanges by the offices of the 189 American bank branches there. Estimates of the dollar volume in trust businesses managed in Nassau ran between $2 and $5 billion. Roughly 95 percent of the trust accounts were for people living abroad.(15)

The I.R.S. turn to The Bahamas started with a pilot trip made by Jaffe and an I.R.S. agent from the Office of Internal Operations (O.I.O.) in December 1962.(16)   In the Registrar General's office in Nassau Jaffe perceived what was happening in The Bahamas. One of the first items he chanced upon was the company file of The Bank of World Commerce which listed all the stockholders, officers and directors of the bank.   Jaffe recognized many of the Bank's officers and stockholders as Las Vegas gamblers and organized crime associates. The information was put into a report which was well received by the Regional Intelligence Office in Atlanta, and I.R.S. Intelligence in Washington.(17) In 1965 the operation, codenamed TRADEWINDS, officially began.

As soon as Tradewinds was approved, a special group of I.R.S. personnel from both the Intelligence Division and O.I.O. went to Nassau to discuss procedures for the Special Agents and to start the process of establishing cooperation and rapport with the U.S. Consul General and his staff and Bahamian police officials.(18) The group conferred with Consul General John Barnard, stating their principal interest was to develop information on American racketeers who have taken "hot money" to The Bahamas and invested it there. Barnard suggested contacting Nigel Morris, Bahamian Commissioner of Police, and Stanley Moir his chief assistant. Others noted as important contacts were Leslie Cate, the principal police officer in Freeport, and Godfrey Higgs, who was, in Barnard's opinion, an honest and knowledgeable lawyer.

The I.R.S. visited all those suggested and found them to be cordial and useful. Additionally, they talked with the Bahamian Comptroller of Customs and the Chief U.S. Customs Inspector at the Nassau airport. They returned to Florida satisfied helpful contacts were established and a sound procedure initiated.   Less than a month later, Jaffe was back in Nassua accompanied by I.R.S. Group Supervisor Richard B. Wallace. They made the rounds talking with their new associates. Assistant Police Commissioner Stanley Moir remarked that there was a serious problem brought about by American swindlers forming Bahamian banks for the purpose of bilking American investors.(19)

From Moir, they paid visits to others cultivated earlier. Eventually, Jaffe ended up at the U.S. Consulate where he met Lieutenant Michael Miller, the Project Liaison Officer for the Navy's important Atlantic Undersea Test and Evaluation Center (AUTEC tested submarines, underwater weapons, and sonar detection systems) on Andros Island. As a result of their conversation, Miller offered to help. He also introduced Jaffe to John Davis, former Assistant Secretary of Agriculture in the Eisenhower Administration. Davis was enthusiastic to serve as an undercover operative and became the first official Tradewinds agent, designated TW-1. Davis and other TW agents were undercover operatives paid for the quality of information received, not merely knowledgeable and helpful individuals.

By the summer of 1965, Tradewinds seemed securely in place. Vital private contacts were established, U.S. Embassy officers consulted, top Bahamian police officials informed, and productive sources engaged. Tradewinds quickly produced around 50 or 60 "information" items of Americans hiding money in Bahamian banks. It was very explosive information, and an increasingly dicey operation made more so by a change in Bahamian law. Prior to November 1965, it was a civil offense in The Bahamas for someone to divulge information protected by the Bank and Company Secrecy Laws; after November, it became a criminal offense. Because of the legal modification, Jaffe and other Tradewinds Special Agents made it a point never to make a direct contact in The Bahamas for information passing. The explicit guidelines for Tradewinds developed subsequently forbade such contacts with Bahamians. Documentary information had to be transferred to an intermediary who would take it to Miami, or the Bahamian sources themselves would bring or mail the material to the I.R.S. in Miami. Sources were paid a fee for information received and had their expenses covered when they delivered material to Miami.

Operation Tradewinds, as a general intelligence-gathering affair, had numerous concerns. For instance, it reported a very disturbing sidelight to the Vietnam War. Late in the summer of 1970, agent TW-19 informed Jaffe that he had been approached by a Washington, D.C. lawyer representing a group with $1 million to invest in land development on Great Exuma Island. The attorney was a former Naval Captain with extensive contacts in the Far East who worked as an assistant to the Navy Department official responsible for the annual procurement of $9 billion of services and hardware.(20) His group was composed of Five U.S. Army Generals serving in Vietnam along with some other U.S. officials stationed there. The lawyer told TW-19 that his organization controlled about $50 million which was sitting in a Hong Kong bank waiting for the right investments. Tradewinds passed this disturbing information up the line of I.R.S. Intelligence where it made its way into the higher reaches of the State Department and then seemed to disappear.(21)

Tradewinds was clearly underfunded--in 1968 the Operation's budget was only $15,000--but still produced substantial money for the government.(22) At the end of 1971, it was estimated that Tradewinds had produced $25 million in taxes and penalties since its inception, and had initiated thirteen full-scale investigations ending in recommendations for criminal proceedings. Overall, it was an inexpensive deal for the government.(23)

ORIGINS OF PROJECT HAVEN

In the spring 1972 Jaffe received a request for aid and information from I.R.S. Intelligence based in San Francisco. The subject was Allan George Palmer (also known as Allan Houseman), born in Pennsylvania in 1940, and involved in the manufacture of LSD and the distribution of it and marijuana in the San Francisco area. He was not a small time dealer. Active since 1968, Palmer was a major producer of LSD, mescaline, and THC (another powerful hallucinogenic drug), and "grass" dealer. It was the Narcotics Traffickers Program, so disliked by Alexander, that had zeroed in on Palmer.(24) As part of the government's drive, Palmer and three others, including an attorney who had clerked for a federal judge, were arrested in Marin County, California, on October 17, 1971.(25)

What brought Jaffe into the Palmer affair were several checks the drug merchant received in May 1970. They were drawn by a Bahamian bank on an account it maintained in the American National Bank and Trust Company of Chicago. The Bahamian bank was identified as the Castle Trust Company, Inc. (The name was changed in 1972 to Castle Bank and Trust Company Ltd.; other slight name changes followed. In common parlance it is called Castle Bank.) The checks worth $22,500 were found in the course of the drug investigation. San Francisco agents requested I.R.S. Intelligence in Chicago and Miami to pursue the money.

The Chicago agents located the Castle account at American National and the signature card on file. There nine names on the card--A. Alipranti; A. R. Bickerton; L. A. Freeman; E. J. Foster; M. S. Gilmour; A. J. T. Gooding; H. M. Wolstencroft; Paul L. E. Helliwell; and Burton W. Kanter--became a kind of centerpiece in the whirling struggle between Jaffe and others from the Intelligence Division, and Alexander and his supporters on the other. The Castle Bank people (owners and some major depositors) strenuously, and for the most part successfully, fought to avoid both criminal and civil actions for tax evasion at the least.

In the beginning, all that the I.R.S. knew about Castle Bank was the Palmer connection, the American National Bank account, and the names on the signature card. Special Agents rapidly learned more. A Chicago agent pulled American National's bank statements enumerating the Castle account. The 1969 and 1970 figures showed Castle maintained a sizeable commercial account; deposits were as large as one million, and several were in the $200,000 to $500,000 range. The account's size aroused I.R.S. interest beyond the issue of Palmer's funds. "We are," wrote the Intelligence Division Supervisor from San Francisco, "concerned about the source of funds being deposited by Castle Trust Company with the American National Bank in Chicago."(26)

The I.R.S. knew much more about Castle Bank by the summer. It was determined Castle had accounts at two other Chicago banks, Continental Illinois National and First National. Moreover, it was discovered that Castle's American National transactions, involving millions in deposits and withdrawals, were handled in three trust accounts. There was also evidence that requests from Castle to Chicago ordering cashier's checks were disbursed to associates of organized crime.(27)

The Service was still tracking details of exactly how Palmer and Castle Bank were introduced. That issue was solved during the early days of the Castle inquiry with information developed by Norman Casper a new Tradewinds operative working for Jaffe.(28) Casper was an experienced private investigator almost 50 years old when he became TW-24. He had past experience as an undercover operative working drug cases for the Miami Organized Crime Strike Force.(29) With Casper's help, Castle was identified as an "underground conduit for untaxed funds to enter banking channels, yet remain undetected," with its own bank accounts in Chicago, South Florida, and New York. Through a Casper source it was learned that Palmer had personally brought some of his money south to a Florida bank called the Perrine-Cutler Ridge Bank for deposit in a Castle account. Most importantly, Jaffe learned Palmer's money never actually left the U.S., the cash stayed in the Castle account at Perrine-Cutler Ridge. Jaffe thus suspected Castle merely provided bookkeeping services, crediting and debiting accounts in which the funds remained in the States. He wrote, "when the depositor requires them again, a debiting entry is made in Nassau with a check or cash paid out by the Miami bank charged only to the CASTLE TRUST account thereby providing the anonymity of the transaction."(30) As time and the investigation of Castle went on, this conclusion that Castle's locus of operation was in the United States was potentially quite significant. If correct (and it was), Castle Bank could hardly be considered a foreign bank and myriad criminal charges might, or should have, followed.(31)

Casper's main source for information was F. (Francis) Eugene Poe, the president of the South Florida Bank (actually two banks--the Bank of Perrine and the Bank of Cutler Ridge which were affiliated through a holding company). In addition to Poe's stateside banking, he was also a Castle Bank director. Casper had known Poe on a casual basis for twenty years, and now approached him under the guise of an undercover drug agent investigating Palmer. The banker believed Casper worked in a private capacity for the federal narcotics bureau and was glad to oblige; he had no idea Casper was with the I.R.S.(32)

Gene Poe gave I.R.S. Intelligence its single greatest investigative opportunity. He paved the way for Casper to actually get inside Castle Bank in Nassau. One day at the Bank of Perrine, Poe introduced Casper to Samuel B. Pierson the relatively young, somewhat unkempt, slightly walleyed, tall, heavy-set president of Castle Bank. (Pierson's presidency, was short-lived, and, it should be noted, was not the key post some might assume. Being president was equivalent to a middle management position, and all policy directives and important decisions flowed directly from American lawyers in Chicago and Miami. Most particularly, the owners and prime movers of Castle Bank were attorneys Burton Kanter from Chicago and Paul Helliwell from Miami). Casper's infiltration went through Poe to Pierson and then to the bank itself. Pierson gave Casper a note introducing him to Castle's vice president in Nassau, Mike Wolstencroft. The message basically said--"Mike, help this guy." Casper was inside Castle by October 1972.(33)

During this autumn visit to Nassau, Casper's penetration paid dividends. He was given a look at Castle's small and often befuddling IBM data processing facility. Among the items Casper saw was an IBM printout several inches thick with lots of corporate and individual names. Left alone for a moment, Casper casually thumbed through it. While doing so, he was genuinely shocked to read the name Richard Nixon on the Castle printout. Later, under trying conditions, Casper stated that he had no idea whether the Richard Nixon entry he saw was actually for President Nixon, or whether someone had put the name there as a joke.(34) At the time, however, Casper believed he had stumbled onto one of the most important investigative leads on political corruption in the history of the U.S.(35) When he reported his findings to Jaffe, he was asked if he could possibly obtain more names? Casper wanted to know if Jaffe would like the printout. Of course, Jaffe replied. From that moment on Casper sought a solution, a way to produce the list.

THE TROUBLED WAY

The method came about through his friendly relationship with Wolstencroft. Since they first met Wolstencroft confided to Casper an interest in extra-marital female companionship.(36) He often travelled and asked Casper to get him dates in Miami.(37) Casper agreed to help. Wolstencroft and other Castle people frequently commuted from Nassau to Chicago to straighten out the bank's confused books in conformity with the Kanter firm, and because they mistrusted the Bahamian government including its postal service. Castle had instituted a policy of hand-carrying documents to Miami and Chicago rather than using the mail.(38)

The opportunity to get the material came in the winter of 1972-73. Wolstencroft planned another Chicago trip, and asked Casper to arrange a date for him in Miami. Such arrangements were routine by then. Casper suggested he see Sybil Kennedy, a woman he had introduced to Wolstencroft a few months earlier. Wolstencroft agreed and called her. The Castle officer had not the slightest idea she was a former policewoman who was now a free-lance operative and sometime journalist. Casper knew Wolstencroft would be carrying the account list and other important Castle documents in his locked leather briefcase when he arrived in Miami on 15 January 1973. Casper shaped a plan. He instructed Kennedy to bring Wolstencroft from the airport to her apartment. On the way, she was to recommend dinner at the Sandbar restaurant on Key Biscayne--a place over at least an hour distant. Casper was confident Wolstencroft would leave the briefcase in the apartment. He notified Jaffe in a memorandum on 9 January 1973, that Wolstencroft would be in Miami on the fifteenth, adding that there was an excellent chance the material could be obtained.(39) Two days before the anticipated event, Casper spoke to Jaffe and gave him the plan in detail. He said he would enter the apartment with a key and the permission of Sybil Kennedy, and take the briefcase out. At that point Jaffe told him to stop; he couldn't and wouldn't authorize this plan without first checking with his superiors and getting their authorization. Later that day, Jaffe told him to proceed; the plan was approved. On the night of the fifteenth, the operation went off without a hitch. Casper grabbed the briefcase and took it to another location where a locksmith made a key to open it, and then to where Jaffe was waiting.   The briefcase was opened and the account list photographed.(40)

The documents were wondrously revealing--the master list of accounts had the names and numbers of over 300 individual and company clients, and other material had the identification and location of around 39 U.S. brokerage accounts for Castle.(41) Jaffe and those involved in this operation were well pleased. They had turned an off-shore bank inside out.

From the moment the film was developed, Jaffe and the I.R.S. Intelligence Division were busy figuring out what they had and how to proceed. This was to be an enormous project and coordination was imperative. The first difficult problem was the identification of the accounts, many of which were in names like Pythagoras, Helwan, Emanon, Topaz, and Inversiones Mixtas. Once that was accomplished, they would have to assign investigations and cases to the appropriate I.R.S. districts. There was a mountain of satisfying work ahead.

In the summer of 1974, Project Haven was moving confidently forward. More and more documents were flowing in and new leads constantly emerging. "We have expanded the scope of the HAVEN Project to include all aspects of worldwide international tax fraud," wrote the project managers. They weren't being immodest. Haven had subsumed many of the functions of two important I.R.S. task forces working on worldwide problems: the "North Atlantic Region Task Force--International Finance" set up to investigate tax fraud and Swiss banks, and the "North Atlantic Region Securities Task Force." They were I.R.S. pioneers moving in what Jaffe described as a virgin, highly specialized area of law enforcement which has been virtually ignored.(42) Their work was organizationally innovative, and the Haven staff provided guidance, documents, and coordination for the

I.R.S. District (Chicago, San Francisco, Reno, New York, etc.) which had many of the Castle cases. It was their most assured period; they felt at the center of an I.R.S. reformation.

VENDETTA

One year later, the Service turned inward and devoured its own. Alexander's move against the Intelligence Division was discernible as early as the winter of 1974-75, although it was the attack on Project Haven and Jaffe which assumed the greatest signficance. Nontheless, the antecedents are important. In the months before the concentration on Haven, Alexander halted the work of an Intelligence Division unit known as the Intelligence Gathering and Retrieval System (I.G.R.S.) which had been accused (inaccurately) of suspicious contact with members of Nixon's White House staff.(43)   The press picked up on the I.G.R.S. story in February 1975, reporting that the Internal Revenue Service was investigating a special Miami unit that gathered information concerning the personal habits and activities of prominent citizens believed to be anti-Nixon.(44)   A few days later the Service denied these allegations, but added it was still investigating. It also announced that Commissioner Alexander's Deputy, William E. Williams, would closely monitor the Miami inquiry. It was later learned that Williams had already started an informal probe of intelligence-gathering practices that past December. This was followed in a month by an announcement that all Intelligence Division activities were to cease except those connected with active tax cases.(45)   And then Alexander and Williams were handed an Intelligence Division nightmare, an operation that seemed to confirm all suspicions of illegality and ineptitude.

"Justice Dept. Is Linked to Sex Spying," announced the startling headline on 16 March 1975. But it was I.R.S. Intelligence not the Justice Department which ran "Operation Leprechaun," a poorly managed affair which grew out of a serious professional probe. Since the early 1970s, there had been a coordinated investigation involving the Miami Strike Force, I.R.S. Intelligence, the Dade County Public Safety Department, and the Miami Police Department, into the association of public officials with organized crime figures in Dade County.(46)

Leprechaun was run by Special Agent John T. Harrison. He worked closely with a veteran informant named Elsa Gutierrez who was previously employed on a casual basis by the Secret Service, Drug Enforcement Administration, and Organized Crime Strike Force. According to Gutierrez, her job was to "get dirt on" about 30 Dade County public officials including three federal judges and a State Attorney who were under surveillance by the I.R.S. agent in charge. Gutierrez implied she was supposed to have sex with at least one of the targets but refused and quit instead.(47) Some of the targets were amused by the attention. "My sex habits are not a taxable item," said a Dade County Circuit Judge. He added a wry note commenting that "if anything, I think I'm depreciating and should get a tax writeoff."(48) Others, naturally enough, were not entertained.

Leprechaun was a perfect foil for slashing away at the Intelligence Division. With its charming name and strange characters, it became a symbol of the Intelligence Division running wild, gone dotty. And though it had absolutely nothing to do with Project Haven, it did have a major impact on its future.(49) Leprechaun formed part of the indispensable backdrop for those committed to stopping Haven. Could the Intelligence Division be trusted? Leprechaun was the answer.

Project Haven was suspended just before it could disseminate Castle Bank file material to the appropriate I.R.S. regions. A meeting in New York was arranged for 19 and 20 August 1975, to provide regional Audit and Intelligence personnel with the files.(50) The Assistant Commissioner, Compliance, concurred with the schedule, but went on record noting that possible adverse publicity might still be generated concerning how the Castle Bank Master List of Accounts and certain other records were acquired.(51) The hedge should have been a tipoff. Within a week or so, Compliance had changed its mind and the meeting was cancelled. The turnabout was peculiarly done, and would provide questions for years of inquiry.

The first version of Project Haven's suspension, told by the highest officials on the I.R.S., held that the decision to suspend Haven was arrived at and carried out by temporary personnel. Edwin P. Trainor, an I.R.S. Regional Commissioner out of Chicago, was detailed the Acting Assistant Commissioner for Compliance for just three weeks in August. The timing was uncanny; the August weeks were crucial for Project Haven's next step. It was Trainor who decided on 13 August to cancel the dissemination meeting and thus to stop the flow of Haven material. His actions were supposedly based on the still unresolved Caseper-Wolstencroft incident.(52) The August decision was an essential element in the imbroglio which erupted within the Service.

Punishing and stopping Jaffe were also necessary and had begun a week earlier than the Haven cancellation. Again the timing is important, for stopping Jaffe and the distribution of Haven documents were two closely related steps in the same policy. But having the Acting Assistant Commissioner execute part two, seemed to give Alexander and his staff some important distance from the issues.

Jaffe's difficulties started during the first week of August with the shock of having his rights read to him by I.R.S. Inspectors sent to conduct an official interrogation. This was an exceedingly serious inquiry, he was told. The matter had already been referred to the Department of Justice for investigation and perhaps prosecution. Jaffe gave them various relevant

memoranda and related that his superior, Troy Register, chief of the
Intelligence Division, Jacksonville District, had been informed of Casper's
plan in advance and approved. After the operation the appropriate I.R.S.
and Department of Justice officials were notified and supplied with the
pertinent details.(53)   Following the interrogation by the Inspectors, Jaffe
contacted Jack Solerwitz, a lawyer employed by the Federal Criminal
Investigators Association (F.C.I.A.), and retained him as counsel.(54)

Solerwitz had recently represented other Special Agents from the
Intelligence Division who were in deep conflict with Alexander. The primary
issue for them was Alexander's policy directive on informants which
constitutes one of the most sensitive factors in professional law enforcement.
Sometime during the first half of May, 1975, Special Agents were ordered to
identify their informants or suffer disciplinary proceedings. Among the many
objectionable features of this policy, reasoned the critics, was the fundamental
one that it put informants at extreme risk. A civil action on behalf of six
Special Agents in New York was filed in the federal court by Solerwitz. The
plaintiffs claimed the policy would wreck their work against organized crime,
drug traffickers, corrupt public officials, and make informing exceptionally
risky and therefore unlikely.(55)

It was impossible to keep this dispute out of the press; a melee of this
magnitude within the I.R.S. was a major story. During the last week of
September, stories about the row monopolized the headlines. Certain details
about Tradewinds, and Jaffe's current predicament became public knowledge.
As expected, the Intelligence Division's dislike and mistrust of Alexander was
one major theme. Sources were quoted saying Alexander's qualifications
were grossly inadequate for the post.(56)   In fact, the press began asking
embarrassing questions of Commissioner Alexander. One report stated there
were allegations that Alexander conspired to undermine a Customs Service
prosecution of one of his former clients, the Proctor & Gamble company.
There were other charges of improper behavior on Alexander's part which
finally forced Treasury Secretary William E. Simon to publicly state he had
complete confidence in Alexander.(57)   After that rhetorical back-slap was
reported, the bad news flowed once again.   Bob Woodward of the
Washington Post noted that Alexander's action had "jeopardized one of the
largest potential criminal and civil court tax recovery operations in the history
of the IRS."(58)

The I.R.S. fight shifted to the Congressional arena when a
Subcommittee of the Committee on Government Operations chaired by
Representative Benjamin S. Rosenthal of New York began an inquiry.
Hearings opened on the morning of 6 October 1975 at the Rayburn House
Office Building. Members of the Subcommittee included Elliott H. Levitas
from Georgia and Father Robert F. Drinan from Massachusetts. Appearing
as a witness the first day was Commissioner Alexander. He was accompanied
by others from his administration involved in the Haven battle, including

Deputy Commissioner Williams, and the <u>Acting</u> Assistant Commissioner for Compliance Edwin Trainor.(59)

Alexander quickly got down to it. He would not stand for uncontrolled law enforcement, for aberrations like Operation Leprechaun, he informed the Subcommittee.(60) The Commissioner was worried about illegally obtained evidence, believing I.R.S. policy on civil cases should match the scrupulousness brought about by the exclusionary rule in criminal matters. Chairman Rosenthal, to the contrary, thought this affair was primarily about the problem of Americans fraudulently bringing money into The Bahamas, deceiving the U.S. Government which was attempting to discover the crimes and collect the taxes.

With that history in mind, Rosenthal noted different undercover methods were used to gather the information, and that these methods generally stayed within the law but sometimes veered a bit to the right or left. The response to his statement was surprising. Deputy Commissioner Williams said the Service wasn't sure about the evidence issue yet; it was still under investigation. With some exasperation, the witnesses were asked how long it took to get the facts about Casper's caper.

Although not mentioned, the Alexander administration already had the facts and at least one legal opinion on the briefcase incident which stated, in effect, the material was useable. Bank secrecy in The Bahamas was not law in the United States. Jaffe had committed no crime.(61) The examination of the briefcase caper was carried out by the Justice Department's Criminal Tax Division in 1975. The primary questions asked were whether the action constituted an illegal search under the Fourth Amendment and an invasion of Wolstencroft's privacy. It was reckoned that the controlling statute which protects the civil rights of both Americans and aliens as to warrantless searches by federal agents is predicated on the action being malicious and without reasonable cause. As far as the briefcase was concerned, the report held, there was plenty of reasonable cause and the search was not malicious. On the most important points, the general conclusions were "no federal criminal statute has been violated through participation in the securing in this country of the records protected by Bahamian law," and the "Bahamian Banks and Trust Companies Regulations Acts of 1965 has not been violated in the course of the briefcase incident, regardless of any question of extra-territorial effect of the Act."(62)

What the Congressional subcommittee was most interested in determining was exactly when Project Haven was suspended, who did it, and why. Alexander was asked, for instance, about press allegations that he had personal reasons both for exposing Operation Tradewinds and closing down Project Haven. The Commissioner denied that was so, adding he "did not shut it [Haven] down."(63) Alexander later strongly reiterated this point which became a theme of the hearing. He could not have served private interests in this affair because he didn't shut Haven down. Ed Trainor made the decision and neither the Commissioner nor the Deputy Commissioner were involved at all. They were only briefed after the fact.(64) He tried to make the Subcommittee understand the Service faced a dilemma. It desired effective, not irresponsible, tax law enforcement. For his commitment to honest and aboveboard enforcement, he said dramatically, "I have been berated by a group of faceless liars."(65)

Some members of the Panel found Alexander's testimony difficult to accept. Congressman Levitas pursued the critical issue of timing, establishing for the record that high officials in both the Justice Department and Internal Revenue Service had known the facts about the briefcase since very early in 1974. Not only that, the Congressmen remarked, but just this past July the question of whether Project Haven should proceed was discussed by officials of the I.R.S., specifically in the Compliance Division, and it was given a green light. Levitas carefully enumerated the troubling sequential issues in the Commissioner's account. Then Father Drinan asked, "What brought everything to a screeching halt?" Deputy Commissioner Williams took a stab at an explanation. It was Operation Leprechaun that made them examine all Intelligence work in the State of Florida.(66)

The chronology of increasing dissatisfaction with the Intelligence Division on the part of the Alexander administration belies this simple relationship--Leprechaun to Jaffe and Project Haven--offered as an explanation. First of all, Deputy Inspector Williams had started an informal inquiry into the intelligence process which resulted in halting Intelligence Division activities on 22 January 1975. Following this an official investigation on information gathering cranked up. Handled by the Inspection Service, the probe focused on the I.G.R.S. in the Jacksonville District. This started on 6 February 1975. It was prompted by the appearance of supposedly damaging stories in the Miami press about I.G.R.S. collecting personal information on prominent individuals. The inquiry thus followed the Deputy Inspector's first negative actions concerning the Intelligence Division.(67)

It was in the midst of this investigation that serious questions about Operation Leprechaun, which had also appeared in the press as early as 27 January 1975, were transformed into an official inquiry. On 14 March, Inspection included Leprechaun in its probe. The unique Operation was brought in because of news reports in March discussing Leprechaun's sex angle. Inspection found Leprechaun was really a term devised by Special Agent Harrison to characterize a network of confidential informants he established in 1972. Working with him was the Miami Strike Force chief who provided the initial targets. individuals legitimately suspected of political corruption.(68)

Inspection's <u>Report</u> on both I.G.R.S. and Leprechaun was completed on 23 June 1975. There were serious flaws evident in I.G.R.S., Inspection asserted, having to do with indexing documents which did not meet the standards of financial transactions or illegal activities with tax potential.(69) The selection system needed work as did the management review of the items indexed. This was the case not only in the Jacksonville District, but also in New York, Los Angeles, and Chicago, areas scrutinized for comparative purposes by Inspection.(70)

Leprechaun's major problems centered on the "free rein" given to Agent Harrison by his supervisors; management's insufficient direction of the operation was the core issue.(71) There were plenty of other suspected irregularities discussed by Inspection having to do with the free-wheeling operation. The last comment offered was that the investigation was continuing and that some matters were already referred to the Department of Justice.(72) Dick Jaffe, Operation Tradewinds, and Project Haven were not mentioned once in the 28 page single-spaced document.(73)

There is a final surprising word on Leprechaun. In the late spring of 1977, Robert W. Rust, U.S. Attorney from the Southern District of Florida, concluded that almost all the goofier examples of I.R.S. bad behavior in this now legendary operation were fabrications. They had either been manufactured or distorted by the Miami News and Elsa Gutierrez for their benefit. Much ado about virtually nothing was his conclusion.(74)

Nevertheless, Leprechaun's disturbing reputation lent a certain plausibility to the claim that all I.R.S. criminal investigations using undercover techniques and informants were necessarily suspect. But it did not lead to Jaffe and his projects. Indeed, if it had, Compliance surely should have been in trouble. It approved the Haven dissemination meeting in July after Inspection's Report was received.

Press reports late in 1975 and early the following year rebutted much of Alexander's testimony at the congressional hearing. The stories went right to the Commissioner's repeated statement that he did not suspend Project Haven. Reporters discovered Alexander was at a meeting and did participate in the decision despite what he and his staff told the Rosenthal Committee. I.R.S. internal documents directly contradicted the Commissioner's testimony. The resolution to suspend Haven came not on 13 August as testified under oath, but the following day. And, although Trainor did authorize the Project's suspension, his action followed a meeting with the Commissioner in his own office (he wasn't out of town after all) at precisely 1:30 in the afternoon on 14 August. One hour later, Trainor telephoned New York I.R.S. officials and told them Haven was to be placed on hold.(75) Trainor finally remembered, the Chicago Sun Times reported, that the action was taken after speaking with Alexander on the fourteenth. Nevertheless, he still claimed he had made his decision the day before and it was accomplished without benefit of the Commissioner's input.

An I.R.S. spokesman tried to explain the damaging account. Alexander "had no advance knowledge of the decision--that is clear. . . . He had advance knowledge of the suspension."(76) The act of suspending Project Haven was now broken down into tiny, willful, independent steps. Alexander and Trainor talked about doing it, but only after Trainor had determined to do it. It wasn't a very clever explanation, but it was the best one fashionable from such slight material.

THE END OF INTELLIGENCE

Alexander broke the back of the Intelligence Division, and understandably some of the victims were rankled. In 1975, for instance, while the Rosenthal Subcommittee worked on the difficult issues, the Federal Criminal Investigators Association sent to the Secretary of the Treasury a Resolution on the havoc caused by Alexander to the Intelligence Division. Included in the Resolution, passed at the society's Detroit Convention in October, was mention of the Strike Force and Narcotic Traffickers Programs disruption, the routine harassment and humiliation of District Directors and managers causing them to further restrict the activities of Special Agents, and the jeopardizing of informants. The F.C.I.A. proposed that the Secretary of the Treasury transfer the criminal investigative functions of the Intelligence Division out of the I.R.S. and place them under the wing of the Assistant Secretary of the Treasury for Enforcement, Operations and Tariff Affairs.(77) The F.C.I.A. had several thousand members and it is hard to imagine a better gauge of broadly held anger and outrage. Almost a year, later certain high ranking Intelligence Division officials from the Southeast Region requested Congressional intervention, claiming an ongoing process of harassment and forced retirement.(78)

Getting rid of Jaffe was policy for several years, but it proved a difficult task. Unlike his Intelligence Division colleagues who retired or backed off under pressure, Jaffe not only stayed in the Service, but remained on Project Haven much to the discomfort of high I.R.S. officials. The Project itself was moved in late November 1975 from I.R.S. control to the Department of Justice and placed under the administration of Senior Oversight Attorney Bernard S. Bailor. Under the direction of the Justice Department, I.R.S. representation was also changed. No longer a National Office Program headquartered in New York, Project Haven responsibilities were, in a sense, reassigned to the Jacksonville District's Miami Field Office. By then, I.R.S. directors and managers in that district reflected the profound changes which had taken place under Alexander's stewardship.

From the moment Haven went to Justice, the Jacksonville managers pressured Bailor to release Jaffe from the Project. Although this wasn't accomplished until the spring of 1977, they were able to marginalize him much earlier. Careful and systematic steps were taken to detach him from the more sensitive Project areas and the decision making process. This was obvious by the summer of 1976, when important planning sessions were held dealing with Castle Bank and Jaffe was not in attendance.(79)

In addition, I.R.S. management prolonged the internal investigation into the briefcase incident which finally ended as Jaffe always predicted. The Inspection Service judged there had been no basis for the charge he had illegally obtained foreign bank documents. Jaffe received an official "clearance letter" from Charles O. DeWitt, District Director, Jacksonville, in mid-December 1976. The new Jacksonville Director regretted any inconvenience the investigation may have caused; the matter was officially closed.(80) But there was still the Justice Department to contend with. The I.R.S. had referred the Jaffe case to Justice for possible prosecution in the summer of 1975. In time, the Department of Justice made exactly the same determination as the I.R.S. A Federal Grand Jury which heard the evidence declined to indict which greatly pleased the U.S. Attorney who was sure Jaffe was the subject of a vendetta.(81)

Despite these findings, I.R.S. leadership still insisted Jaffe had to go. Finally a way was found which corresponded in time with the first major Castle Bank depositor case. Charged in this important action was Jack Payner from Cleveland who failed to report his Castle Bank account which held $442,000. Ironically, Payner's indictment was based in part on his failure to honestly answer the foreign bank account question which so bothered Alexander he successfully fought to have it removed from the form. Because the government's evidence was linked to Wolstencroft's briefcase documents, Jaffe was an important witness. A pattern of new harassment emerged at this time, designed to take the marginal Jaffe and banish him forever from Project Haven.

In preparation for suspending him, a series of niggling charges were filed. Bernard Bailor complained to the I.R.S. Inspection Service that Jaffe's testimony at the Payner trial conflicted with earlier sworn statements made to the Inspection Service. A Special Inquiry was therefore begun to consider the charge on 25 March. The investigation lasted until almost the end of May, neatly bracketing much of the trial period. At its conclusion, Jaffe's testimony was deemed consistent.(82) However, Jaffe's problems weren't over with. He was also charged with disclosing Project Haven information to a private attorney. Almost a year after the Payner case, DeWitt was again pleased to inform him the allegation was groundless and he regretted any inconvenience.(83)

The decision to remove Jaffe was made early in April 1977, and a plan put into motion. Support from the Justice Department was very important, and I.R.S. officials worked diligently to secure it. The forces against Jaffe were in line no later than 22 April.(84) They were waiting for the most propitious moment. The Payner case provided it.

On a motion to suppress the evidence, Federal Judge John M. Manos, in the Cleveland District Court, levelled a staggering judgement on Jaffe and his actions. Judge Manos, on 28 April 1977, found Jaffe's testimony on certain points "unconvincing," and the "search procedure" outrageous "shocking the court's conscience." He decided Casper's caper was "inconsistent with American standards of justice," and that Jaffe's and Casper's conduct was illegal. Manos thus held the government's evidence was the fruit of an illegal seizure of Wolstencroft's briefcase, and had to be suppressed.(85) The next day, "in light of the Payner decision" he was told, Jaffe was off Project Haven.(86)

The die had already been cast, but the absolutely withering opinion made the job infinitely easier. Jaffe was not surprised by his suspension, although he suspected and came to know that the Payner case covered rather than initiated the action.

Several influential columnists and reporters commented on the case. Jack Anderson, for one, defended Jaffe and lambasted I.R.S. officials. Anderson said the Internal Revenue Service jeopardized over a $500 million investigation of rich taxpayers by removing its premier agent from the case. He added, this could only be good news for mobsters. The columnist pointed out the Manos decision was contrary to the findings of a federal Grand Jury and an 18-month investigation conducted by the I.R.S..(87)

Jaffe retired from the Internal Revenue Service in January 1979. About a year later, the Payner case reached the Supreme Court. It was argued in February 1980, and decided in June.(88) The Manos opinion was rejected by a vote of six to three. Chief Justice Burger wrote "Payner--whose guilt is not in doubt--cannot take advantage of the Government's violation of the constitutional rights of Wolstencroft, for he is not a party to this case."(89) In this vastly imperfect world, the fruits of Casper's caper were ultimately determined adequate.

ALEXANDER'S LEGACY

The termination of Tradewinds and the battle over Haven, shut the only window American law enforcement had into American criminals using The Bahamas. This was apparent as early as the October 1975 congressional hearings when Alexander was asked questions about issues beyond Project Haven. Of particular concern was the Commissioner's policy for dealing with the apprehension of tax evaders using off-shore havens.

In an attempt to summarize the testimony on recommended procedures for dealing with countries such as The Bahamas, Representative Anthony Moffett of Connecticut noted a counter-productive logic. He argued if a country such as The Bahamas passes bank secrecy legislation, and if the

I.R.S. will not use any information on tax fraud that violates a foreign law (Alexander's position), wasn't this, therefore, a policy declaring that off-shore tax evasion in countries with strict bank secrecy laws was safe from the Service? There was no satisfactory answer to Moffett's analysis. The best the Commissioner could offer was the need to pursue tax fraud issues through proper diplomatic channels.

Then the Commissioner was questioned about the degree of past cooperation on tax and banking matters by the Bahamian government. There was an awkward and embarrassing exchange. Neither Alexander nor his staff at the hearing had a clue concerning past Bahamian/U.S. cooperation in tax matters that came through diplomatic channels. The only historical examples of cooperation they could have cited were those forged by Jaffe with the Bahamian police, and expressed through the years in Operation Tradewinds.

> Mr. ROSENTHAL. Has the Bahamian Government been cooperative with our Government in the enforcement of the laws of the United States: Does anybody have an opinion?

> Mr. ALEXANDER. Does anybody from IRS back there have any thoughts on this? Mr. DeWitt [District Director from Jacksonville]? Do you have any thoughts on this?

> Mr. DeWITT. I cannot speak to the period we are covering.

> Mr. ALEXANDER. I am sorry we cannot help you, Mr. Chairman, but we look forward to another session in which we may be able to give you some help.(90)

By the end of the decade drug smugglers, organized crime syndicates, and tax evaders, shielded from American law enforcement scrutiny and in cahoots with corrupt Bahamian government officials, had turned The Bahamas into a racketeer's paradise. A fair share of the responsibility for this had to be placed at Alexander's door. This was done, in fact, by David R. MacDonald who had been the Assistant Secretary for Enforcement, Operations and Tariff Affairs at the Treasury Department in 1974.(91) He supervised the law enforcement activities of several Treasury Department branches such as Customs, the Bureau of Alcohol, Tobacco and Firearms, and the Secret Service.

Testifying before Congress in 1980, MacDonald maintained that the adoption of Alexander's policy forced the Service to engage in reverse discrimination by concentrating on the returns of wage earners instead of hard-core criminals. Smugglers and other professional criminals, he told a Congressional audience, learn to isolate themselves from the proscribed goods and services they sell and thereby avoid prosecution. But they cannot and would not separate themselves from the money which was precisely why they went into the business of crime. He concluded that the I.R.S. was indispensable in any earnest effort against the international drug traffic and it had to abandon the lenient, discriminating, and foolish policies of the past.

There was no legitimate reason for dropping the narcotics program and ending cooperative work aimed at organized crime. The attack on the Intelligence Division didn't advance civil liberties even by happenstance. It merely took advantage of the national uproar to supply an "after the fact" justification for a policy that made the world safer for international organized crime and its helpers in law and banking. There never was any necessity to end Operation Tradewinds nor to treat Haven in the manner that was done. Whatever procedural mistakes were made could have been fairly treated far short of the draconian measures used. But any deviations were grist for the mill built to crush I.R.S. Intelligence, likely for its audacity, surely not for its misdemeanors.

* * * * * * * * * * * * * * * * * * *

NOTES

1. Alexander was born in Pine Bluff, Arkansas, in 1921. He married in 1946 and had two sons. Alexander's education was notable; a B.A. from Yale in 1942 (with honors), and an L.L.B. from Harvard (magna cum laude) in 1948. In Between Yale and Harvard, Alexander served in the Army with the 14th Armored and 45th Infantry Divisions. He was awarded Silver and Bronze Stars. After the war he worked for a short while with a Washington, D.C. law firm, and then moved to Cincinatti where he became in 1968 a senior partner with the well-known firm of Dinsmore, Shohl, Coates & Dupree.

2. The Intelligence Division when first formed in the Bureau of Internal Revenue (as the IRS was then called) was known as the Special Intelligence Unit. Hank Messick, Secret File (G. P. Putnam's Sons, 1969), 22-23.

3. See the Internal Revenue Service, Annual Report (Government Printing Office, 1973), 96, 97.

4. In 1965 the IRS employed around 60,000 people in various offices and posts. There were two kinds of Agent--12,718 Revenue Agents working for the Audit Division, and 1,712 Special Agents in the Intelligence Division. IRS, Annual Report (1966), 68.

5. See the story about Alexander's tax philosophy in U. S. News and & World Report, 27 February 1976.

6. IRS, Annual Report (1973), 30.

7. Clark R. Mollenhoff, Strike Force: Organized Crime and the Government (Prentice Hall, 1972), 132.

8. IRS, 31.

9. U.S. House of Representatives, Subcommittee of the Committee on Government Operations, Oversight Hearings Into The Operation Of The IRS: Operation Tradewinds, Project Haven, And Narcotics Traffickers Tax Program (Government Printing Office, 1976), 4-5, 1267.

10. See U.S. House of Representative, Committee on Government Operations, INTERNAL REVENUE SERVICE AND TREASURY DEPARTMENT ENFORCEMENT OF THE FOREIGN BANK ACCOUNT REPORTING REQUIREMENT OF THE BANK SECRECY ACT Second Report (Government Printing Office, 1977).

11. U.S. Senate, Permanent Subcommittee on Investigations of the Committee on Government Operations, Hearings: ORGANIZED CRIME, Securities: Thefts and Frauds (Second Series), 93rd Congress, First Session, part 1, 29 June, 13 July 1973; also see Permanent Subcommittee on Investigations, Hearings: ORGANIZED CRIME, Stolen Securities, 92nd Congress, First Session, part 1, 8, 9, 10, 16 June 1971 and part 3, 20, 21, 22, 27, 28, 1971 (Government Printing Office, 1971).

12. David Wise, The American Police State: The Government Against the People (Vintage, 1978), 322-351.

13. The Bahamian project was completely involved with Dick Jaffe's professional life. He was its inspiration and prime architect. It is literally impossible to follow the contours of the I.R.S. effort in The Bahamas outside the context of Jaffe's career. It was neither easy nor simple to move the I.R.S. to support Jaffe's investigative inspiration as there was internal opposition almost from its inception. If not for jaffe's doggedness, the operation would have quickly faltered. But he stubbornly persisted and it didn't although there was a future heavy price he would pay for his initial success. Jaffe was born and raised in Brooklyn, New York, and moved to Miami in 1948. Tall, balding, soft-spoken and articulate, he went to the

University of Miami where he studied business administration, marketing and advertising, graduating in 1950. Less than a year later, he was in the Army heading for Korea. A forward artillery observer in the 7th Infanty Division, Second Lieutenant Jaffe fought in the famous battles of Old Baldy and Pork Chop Hill in the waning months of the war. He returned home in 1953 as a First Lieutenant. Jaffe returned to school, this time earning a degree in accounting. A few years later, because of a friend's prodding, he took the I.R.S. Civil Service exam (scored the highest mark in the Southeast Region) and joined the Service as a Special Agent with a yearly salary of $4,525.

Jaffe investigated criminal violations of revenue statutes including the wagering tax law designed to ferret out bookmakers and lottery operators. In 1957, he was assigned several cases involving small-time gamblers on the periphery of major organized crime syndicates. The following year, he spent 22 weeks in Washington at Treasury's Law Enforcement Officer Training School where he graduated first in his class. Back in Miami, he continued working on routine investigations until 1960 when he was assigned to an important organized crime case. It was this investigation which sparked Jaffe's interest in pursuing racketeers in The Bahamas. The case dealt with the finances of Michael "Trigger Mike" Coppola identified as a high ranking member of the Genovese organized crime syndicate, one of New York's most powerful mobs.

14. The Bahamian off-shore banking industry grew rapidly as a glance at Rand McNally International's Bankers Directory shows. In 1962 there were very few private banks or trusts listed in the Directory's Bahamian section. Banking was carried out by the branches and/or subsidiaries of familiar firms such as Chase Manhattan, Barclays, the Royal Bank of Canada, the Bank of Nova Scotia, the Bank of London and Montreal, and New York's First National City. That year Bahamian banking institutions were comfortably listed on a single page in the Directory. Tens years later, the banking scene had changed dramatically. The 1962 institutions were still around, but their assets, as well as branches and subsidiaries boomed. The 1972 list covered more than five pages in the Directory. The American banking industry was well represented with Bahamian counterparts of banks headquartered in Boston, Philadelphia, Cleveland, New York, Chicago, Dallas, Seattle, and Grand Rapids Michigan.

15. Richard H. Blum, Offshore Haven Banks, Trusts, and Companies: The Business of Crime in the Euromarket (Praeger, 1984), 135.

16. The following material on the origins of Operation Tradewinds is derived from taped and transcribed interviews with Richard Jaffe.

There was conflict between Jaffe and the O.I.O. agent because of significant differences over the function of law enforcement between the two I.R.S. divisions. The O.I.O. had primary responsibility for overseas intelligence gathering, but it had little or no interest in that kind of work. O.I.O. leadership resolutely opposed overseas investigations and that meant there would be an inevitable clash with the Intelligence Division. The O.I.O. always attempted to stop Intelligence Division Special Agents from going abroad on the grounds that their kind of work caused diplomatic problems.

17. However, the report encountered stiff opposition from the O.I.O. section which claimed his report was inflated, insignificant and misleading. Subsequently, the O.I.O. stubbornly fought an Intelligence Division recommendation to establish a Bahamian operation to be run by Jaffe. One part of the recommendation called for the operation to receive an exemption from O.I.O.'s offshore jurisdiction in order to work The Bahamas without interference. O.I.O. battled the Intelligence Division for over two years before the recommendation was finally approved. Jaffe did not patiently wait the two years it took for his operation to be institutionalized. With the approval of his superiors in the Jacksonville District he began his activities earlier. Reports he wrote from material gathered in The Bahamas were enthusiastically passed to Washington by Troy Register Chief of the Intelligence Division in Jacksonville, and Jim Sutherland, the Regional Organized Crime Coordinator in Atlanta. It was the pressure brought by the Intelligence Division armed with Jaffe's "extramural" reports which finally overcame O.I.O. opposition and secured the project's endorsement.

18. Memo to File, Richard B. Wallace, Re: Official Visit to Nassau, B.W.I., 4 and 5 March 1965.

19. Richard E. Jaffe, Report, 1-2 April 1965.

20. Director, Intelligence Division, Washington, D.C. to Regional Commissioner, Southeast Region, "OPERATION 'TRADE WINDS,'" 27 October 1970.

21. Turner B. Shelton letter to Honorable Edwin W. Martin, American Consul General, American Consulate General, Hong Kong, 28 August 1970.

22. D.W. Bacon, Assistant Commissioner (Compliance) to William J. Bookholt, Regional Commissioner, Southeast Region, "Confidential Expenditures, Jacksonville District, Southeast Region, OPERATION 'TRADE WINDS,'" 9 September 1968.

23. J.A. McRae, Group Supervisor, Miami III, to Chief, Intelligence Division, Jacksonville, "Operation Tradewinds Work Plan--FY 1972," 13 August 1971.

24. Paul H. Wall, Group Supervisor, Intelligence Division, San Francisco District to District Director, IRS, Attn: Chief, Intelligence Division, Jacksonville, Florida, <u>MEMORANDUM Subject: Allan G. Palmer 94-22-179-4-7 (NTP)</u> 22 May 1972.

25. "Half Ton of 'Pot' Jails Four," San Francisco Examiner, 18 October 1971, 3.

26. Ibid.

27. Intelligence Division (Miami)to District Director of Internal Revenue, Attn: Chief, Intelligence Division Jacksonville District, "Memorandum: Joint Compliance Program, Quarterly Narrative Report," 7 July 1972.

28. Intelligence Division (Miami) to Chief Intelligence Division, San Francisco District, "Collateral Reply," 27 June 1972.

29. See the Chairperson, Judicial Council of California, Judicial Coordination Proceeding No. 1040, San Francisco Superior Court No. 764340, Santa Barbara Superior Court No. 121771, "Musical Group Investment Cases, Deposition summary of Norman L. Casper," Vol. 1, 14 April 1982.

   I received my copy from papers on file at Bishop, Barry, Howe & Reid, Attorneys-At-Law, 220 Bush Street, San Francisco, California.

30. State of Florida, County of Dade, "AFFIDAVIT of Richard E. Jaffe," 8 February 1978, 2.

31. Castle people were not fools and as soon as the I.R.S. served a summons on the Chicago bank for its Castle account records, there was a response in Nassau. They applied to the Bahamian government for a name change and most importantly a general banking license which would legitimize transactions for outsiders that were clearly and patently illegal under Castle's existing Bahamian license. See Intelligence Division (Miami) to District Director of Internal Revenue, Chief Intelligence Division, Jacksonville district, "Joint Compliance Program, Quarterly Narrative Report," 7 July 1972.

   They got the name change--from Castle Trust to Castle Bank & Trust Ltd.--but not the desired unrestricted banking license. Their working license had been issued on May 12, 1971, and had the following restrictions: "a) it carries on no banking business with the public; and b) it only carries on banking business of its shareholders and persons, firms or corporations affiliated with or controlled by its shareholders." TW-5, "Report on Castle," 23 June 1972. Castle knew that restriction B could prove very troublesome if ever their depositors became known.

32. Musical Group Investment Cases, "Deposition Summary of Norman L. Casper."

   The evidence of Castle's criminality was growing and there were I.R.S. meetings held during which investigative coordination was planned. Liaison with the Department of Justice Tax and Organized Crime Divisions was discussed. Late that summer the the I.R.S. summarized what it had: twenty-one identified Castle directors, officers and shareholders; over half a dozen Castle commercial accounts in American banks; several organized crime associations and four narcotics targets; many prominent U.S. citizens involved in stock transfers to Castle which then acted as their trustee; and six American brokerage and investment companies which paid Castle dividend income. The I.R.S. intended to start a grand jury investigation in the very near future. See Norman J. Mueller, "Summary Report, Castle Trust Company," 5 September 1972.

33. TW-24, "Memorandum to File, Re: Castle Bank & Trust Company," 5 September 1972.

34. U.S. House, 157-158.

35. U.S. House, 179.

36. The compact Scot, born in 1932, educated at Trinity College, Glenalmond, Scotland, and years later at Sir George Williams University in Montreal, Canada, where he earned a Bachelor of Commerce degree with a Major in General Administration, was married and had a ten-year old daughter at the time of his request to Casper. See Herbert Michael Wolstencroft, "CURRICULUM VITAE," in author's possession.

37. U.S. House, 159; and TW-24, "Memo to File . . ." 13 October 1972, 3; and, "Affidavit of H. M Wolstencroft" taken by the I.R.S. in the Commonwealth of The Bahamas, 13 January 1976.

38. U.S. House, 160; and Musical Group Investment Cases, "Deposition Summary of Norman L. Casper," 5.

39. U.S. House, 177.

40. Musical Group Investment Cases, "Deposition Summary of Norman L. Casper," 7; House of Representatives, 178.

41. Intelligence Division (Miami) to Chief, Intelligence Division, Jacksonville District, "Castle Trust, et al and Operation DECODE, Briefing Summary," 29 March 1973.

42. Field Project Managers to Assistant Regional Commissioners, North Atlantic Region, "Project HAVEN Current Status and Future Plans," 11 November 1974.

43. Arnold Markowitz, "IRS: We Halted All Intelligence Except on Crime," Miami Herald, 29 January 1975.

44. Andy Rosenblatt, "IRS Spy-for-Nixon Report Spurs Quiz," Miami Herald 2 February 1975.

45. Arnold Markowitz, "IRS Finds No Misuse of Data," Miami Herald 5 February 1975.

46. Dougald D. McMillen, Chief Miami Strike Force to William S. Lynch, Chief Organized Crime and Racketeering Section, "MEMORANDUM: Sensitive Case Report," 23 May 1974, 1; and McMillen to Lynch, "MEMORANDUM: Organized Crime--Corruption Investigation Dade County, Florida," 9 September 1974, 3.

47. Mike Baxter, "Sex Spy Reveals Identity: 'IRS Agent Threatened Me,'" Miami Herald 16 March 1975; and William R. Amlong and Gene Miller, "Woman: IRS Probed Dade Sex Lives," Miami Herald 15 March 1975.

48. Ron Sachs and Steve Strasser, "Anger and Humor Greet IRS Report," Miami Herald 16 March 1975.

49. A conclusion reached by former Department of Justice Attorney William D. Hyatt who worked Project Haven through the Leprechaun period. Author's interview 7 July 1987.

50. Regional Commissioner, North-Atlantic Region to All Regional Commissioner . . . "Memorandum: Project HAVEN," 22 July 1975.

51. S. (Sing) B. Wolfe, Assistant Commissioner (Compliance) to Regional Commissioner, North-Atlantic Region, "Memorandum: Project HAVEN," 16 July 1975.

52. E.P. Trainor, "SUMMARY OF FACTS SURROUNDING THE DECISION TO DEFER DISSEMINATION OF LEADS CONCERNING OPERATION TRADEWINDS-HAVEN," 3 November 1975.

53. Richard E. Jaffe, "MEMORANDUM OF INTERVIEW," 7 August 1975. Guilford Troy Register Jr. supported most of Jaffe's recollection when he testified several years later during one of the few Haven cases. Register remembered he took the copied briefcase documents to Atlanta, and presented the material both to the Regional Commissioner of Internal Revenue for the Southeast Region and the Assistant Regional Commissioner for Intelligence, the head of Intelligence for all the Southeast. From Atlanta, Register transported the documents to Washington and turned them over to the Director of Intelligence, John Olszewski.

Register and Jaffe disagreed over the question of prior knowledge of the briefcase incident. The former District Intelligence chief remembered a telephone conversation with Jaffe asking for permission to receive and photograph documents brought by an informant. But he recalled only a brief talk during which no mention was made of how the informant gathered or intended to gather the material. In any case, he authorized the receipt and reproduction of the documents. Even when he did learn the details, Register didn't pause for a moment in moving the project forward. See United States District Court, Southern District of Florida, Magistrate Division, USA vs. James M. Moran, et. al., Case No. 78-189-Cr-ALH, "Testimony of Guilford Troy Register, Jr.," 12 February 1981, 51-53.

54. Richard E. Jaffe to Mr. Jack Solerwitz, "Letter," 26 August 1975.

55. U.S. District Court, Eastern District of New York, Frank J. Scodari, et. al., Plaintiffs against Donald C. Alexander, Commissioner of the Internal Revenue Service and William E. Simon, Secretary of the Treasury, Defendants, Civil Action No. 75C813.

56. Charles Osolin, "Feud Between IRS Director and Federal Law Enforcers," Cox Newspapers, Washington Bureau, 27 September 1975.

57. James Savage and Phil Gailey, "IRS Chief Faces Inquiry in Probe of Bank Accounts," Miami Herald 27 September 1975.

58. Bob Woodward, "IRS Shelves Inquiry Into Tax Havens," Washington Post, 27 September 1975.

The material in the American press jarred Bahamian sensibilities in several ways. One particular story in the Miami Herald produced a corresponding answer from The Bahamas. The Miami account reported that Jaffe had confidential informants in The Bahamas who were in top political and financial circles in Freeport and Nassau. The following day the Bahamian government announced it had known of Tradewinds since 1972, considered it an attempt by the I.R.S. to breach the country's bank secrecy laws, and had sacked an immigration officer for aiding the operation. See Nicki Kelly, "Immigration man sacked over IRS link, Nassau Tribune 24 September 1975.

Secretary of State Henry Kissinger instructed the Ambassador in Nassau, if questioned by Bahamian officials, to respond that the Service was conducting an internal investigation of Tradewinds. See SECSTATE WASHDC to AMEMBASSY NASSAU PRIORITY, "SUBJECT: IRS PROBE IN BAHAMAS," September 1975. By the end of September, tempers in The Bahamas were frayed over the revelation of Tradewinds. At one point, the Prime Minister threatened to punch out the few teeth of an independent representative in the House of Assembly, who had the temerity to call for an investigation of foreigners using The Bahamas for illegal purposes. The Prime Minister shouted that the representative himself might be involved with an I.R.S. informant. Joining the Prime Minister in his attack was the Deputy Prime Minister who threw a book at the beleaguered representative. American Embassy Bahamas to SECSTATE WASHDC, Amembassy LONDON, "SUBJ: IRS PROBE INTO BAHAMAS," CONFIDENTIAL NASSAU 1628, 25 September 1975.

59. U.S. House, 2.

60. U.S. House, 20-21.

61. David E. Gaston, Director, Criminal Tax Division, to Meade Whitaker, Chief Counsel, "Internal Revenue Service MEMORANDUM: Draft Tradewinds Report dated July 3, 1975," 24 July 1975, 2.

62. Gaston, 3.

63. U.S. House, 30.

64. U.S. House, 53.

65. U.S. House, 32.

66. U.S. House, 63.

67. I.R.S., Inspection, 1.

68. I.R.S., 23.

69. I.R.S., 5.

70. I.R.S., 6.

71. I.R.S., 20.

72. I.R.S., 28.

73. Robert W. Rust, United States Attorney, to the Honorable Griffin B. Bell, Attorney General, and Honorable Jerome Kurtz, Commissioner Internal Revenue Service, "Re: OPERATION HAVEN, OPERATION TRADEWIND," 17 June 1977.

This was the same judgement reached earlier by a Detective from a small Broward County police department who knew Special Agent Harrison, worked with the Miami Strike Force, and wrote President Gerald Ford. The Detective informed the President that Alexander was out to destroy investigations of organized crime and corruption. He implored the President to remove the Commissioner described as an "incompetent bumbling nincompoop." See Detective Lieutenant Pierre Pelletier, Intelligence, Oakland Park Police Department to Mr. Gerald R. Ford, President of the

United States, "Re: John Thomas Harrison and Operation Leprechaun," 24 April 1975.

74. Robert W. Rust, United States Attorney, to the Honorable Griffin B. Bell, Attorney General, and Honorable Jerome Kurtz, Commissioner Internal Revenue Service, "Re: OPERATION HAVE, OPERATION TRADEWINDS," 17 June 1977.

75. U.S. House, 1269-1270.

76. U.S. House, 1268-1269.

77. Fred C. Denham, National President FCIA to Hon. William Simon, Secretary of the Treasury, "FCIA RESOLUTION," 20 October 1975.

78. The petitioners were Ed Vitkus, Assistant Regional Commissioner of the Intelligence Division, Atlanta; A. J. O'Donnell, Jr., Regional Commissioner of the Southeast Region, Atlanta; and Troy Register, Intelligence chief in Jacksonville. See E. J. Vitkus, Assistant Regional Commissioner (Intelligence), Southeast Region to the Honorable Charles A. Vanik, Chairman, Subcommittee on Oversight, Committee on Ways and Means, U.S. House of Representatives, "Letter," 8 March 1976, and A. J. O'Donnell, Jr., to the Honorable Charles A. Vanik, Chairman, Subcommittee on Oversight, Committee on Ways and Means, U.S. House of Representatives, "Letter," 4 March 1976.

79. James J. Lane, Group Manager, Jacksonville District, "MEMORANDUM OF MEETING" 30 and 31 August and 1 September 1976.

80. Charles O. DeWitt, District Director, Jacksonville, to Richard E. Jaffe, Special Agent, Intelligence Division, 20 December 1976.

81. See Rust, "Re: OPERATION HAVEN."

82. Jeffrey J. Motyka, REPORT OF INVESTIGATION, 13 June 1977.

83. Charles O. DeWitt, District Director, to Richard E. Jaffe, Special Agent, 28 March 1978.

84. Anthony V. Langone, ARC (Intelligence Southeast Region), "MEMORANDUM TO THE FILE," 3 May 1977.

85. U.S. District Court for the Northern District of Ohio, Eastern Division, USA v. Jack Payner, Case No. CR76-305, Judge John M. Manos, MEMORANDUM OF OPINION MOTION TO SUPPRESS 28 April 1977.

86. James J. Lane, Acting Branch Chief, Miami, Florida, to Special Agent Richard Jaffe, Group 902, Miami, "MEMORANDUM: PROJECT HAVEN," 29 April 1977.

87. Jack Anderson, "IRS Pulls the Rug Out From Top Miami Sleuth," 14 May 1977.

88. U.S. Supreme Court Reports, United States v. Jack Payner 65 L Ed 2d (1980), 468-487.

89. U.S. Supreme Court Reports, 478.

90. U.S. House, 86.

91. See U.S. Senate, Permanent Subcommittee on Investigations, Report: Illegal Narcotics Profits, "Statement of David R. MacDonald," 4 August 1980.

Chapter 10
Anti-communism and the War on Drugs
Published in <u>Contemporary Crises</u> (1990).

During the past decade a rhetorical "war on drugs" has taken the nation by storm. Hardly a politician can be found willing to consider any but the most severe punishments for drug dealers and users. "Zero tolerance" and calls for stiffer penalties are calculated to give the impression that the government will give drug merchants no quarter. It appears that a national panic has been manufactured; after months of television news proclaiming that drugs are the nation's number one problem, the story is confirmed by television polls asking television viewers what they think the nation's number one problem is. A less scientific more bogus effort is difficult to imagine. In this environment suggesting that this Administration is using the drug issue to militarize American foreign policy in Latin America, is to invite ridicule. Yet there are now more than 30 U.S. government entities involved in the war on drugs in Latin America alone, including the DEA, U.S. Customs, the U.S. Information Agency, The Bureau of International Narcotic Matters, the FBI, the CIA, the Agency for International Development and a host of others.(1) In my view, their overall purpose is to implement traditional hemispheric American policies, using the pretext of drug trafficking to justify this remarkable paramilitary and intelligence presence.(2) Citing the cocaine menace, the U.S. has done overtly what it used to do only covertly--influence Latin governments, especially their military and police.

Recently, for instance, United States foreign aid allocations have been tied to cooperation with drug enforcement efforts, no matter how demanding and unrealistic these demands might be. Furthermore the proportion of such aid that is earmakred for drug enforcement has increased dramatically. The amount of total aid to Colombia required to be spent on drug interdiction grew from 30 to 70 percent between 1984 and 1987.(3) These requirements include close collaboration with U.S. drug and intelligence organizations in the destruction of cocaine production facilities. Recipient nations must also develop and arm special police and military units for joint drug interdiction efforts under the control of U.S. authorities. Although these are generally described as cooperative, bilateral agreements with American personnel acting in advisory capacities, they are actually unilateral demands to be carried out under American supervision.

Historical Antecedents

The core of American drug policy lies within the traditional concerns of U.S. foreign policy, and that has not changed in many decades. It consists of standard fare anti-communism; this explains why the U.S. has generously supported the Afghan rebels in their fight with the Soviet Union, despite the fact that they are among western Asia's most notorious drug smugglers. This situation is, of course, reminiscent of the American "presence" in Southeast Asia during the Vietnam War, when U.S. Special Forces flew Golden Triangle opium and heroin to Bangkok and other centers for distribution, and supported various hill tribes whose economy depended exclusively upon opium production. These groups received American assistance as long as they remained stalwarts against Vietnamese communism.(4) Despite the rhetoric about our nobility of purpose in Southeast Asia, the plain fact is that we favored organized criminals in the drug trade as well as vice, currency, precious metals, and extortion rackets.

Scholarship by by Frances Fitzgerald, Alfred McCoy and William Corson published two and three decades ago, demonstrated the subordination of drug control to anti-Communism.(5) Corson, who was a Marine Intelligence officer in Vietnam before he took up scholarship, understood the relationship perfectly. That is why he blew up a U.S. forces clandestine narcotics factory in Laos, to show his displeasure with the structure of foreign policy.(6)

Even before our formal involvement in the Vietnamese conflict, American CIA officers were instrumental in promoting the region's narcotic trade. Again, the "quid pro quo" was support against the Communist menace. In this regard the actions and activities of banker, lawyer, and intelligence officer, Paul Lionel Edward Helliwell, are instructive. During World War II, Helliwell served in Army's G-2 Intelligence Group in the Middle East and then transferred to the Office of Strategic Services (OSS) as Chief of Intelligence in China commanding 350 Army and Navy personnel.(7) It was Helliwell and members of his OSS group who had contact with Ho Chi Minh during the latter days of the war. Ho and Helliwell had three meetings between January and the end of March 1945. At each encounter Ho requested arms and ammunition to fight the Japanese; Helliwell refused unless Ho would give categorical assurances that the weapons would not be used against the despised French. They deadlocked over the issue. When the OSS disbanded at the end of the war, Helliwell became Chief of the Far East Division of the Strategic Services Unit (SSU) of the War Department until the spring of 1946. The SSU was the interim intelligence unit bridging the gap between the dismantling of the OSS and the creation of the CIA in 1947.(8)

After Helliwell was mustered out of the military he joined a small Miami law firm specializing in real property, insurance, tax, trade regulation, and so forth.(9) He also worked for the CIA. Initially, his activities centered on Southeast Asian problems as he joined with the famous and controversial Major General Claire Chennault, known as an "acerbic warrior, at odds with his superiors for decades."(10)

Chennault was one of the first exponents of the Southeast Asian "domino theory" of Communist expansion. And he had a plan to stop this imagined catastrophe. In his vision, through the skillful use of air power supplying war material to more-or-less indigenous anti-Communist Chinese, coupled with the employment of American military advisers for training and planning, Chinese communism could be contained, thus saving all of Southeast Asia. the necessary logistical support for these operations would be furnished by certain civilian airlines, such as the Civil Air Transport (CAT), owned as it happened by Chennault and a partner.(11)

There were few takers in Washington for Chennault's plan even though he had the lobbying help of attorney Thomas G. "Tommy the Cork" Corcoran, an early FDR brain-truster, zealously committed to New Deal policies until he discovered how lucrative the other side was. Corcoran then became known for sleazy influence peddling, lobbying and backdoor deals.(12) Chennault was stymied until his friend Paul Helliwell intervened. It was Helliwell who broke the impasse by suggesting to Frank Wisner, an important CIA official, that he use CAT for Southeast Asian operations. The powerful and well-placed Wisner was at the time in charge of the Office of Policy Coordination, a covert action organization tied to both the CIA and Department of State in 1948-49, but soon entirely within the Agency. Wisner agreed, and requested that Helliwell figure out a clandestine way to subsidize the airline which was then in desperate financial trouble.(13)

In the autumn of 1949, a formal agreement was reached and signed by Corcoran for CAT and a CIA representative from the Office of Finance.(14) Helliwell helped construct the CIA's commercial cover organization for CAT and its Southeast Asian covert missions. This was the Sea Supply Corporation set up in Florida with its main office located in Bangkok, Thailand.(15) In 1952 Helliwell's law firm became General Counsel for Sea Supply. That same year Helliwell was made the General Counsel for the Royal Consultate of Thailand.(16)

Under the stewardship of Helliwell, Sea Supply did far more than cover the CAT operation. It channelled assistance to the Thai Chief of Police who was deeply implicated in the region's extensive opium trade. With Sea Supply's help, the Chief built a police force of 42,833 men which rivalled, and most likely surpassed, the Thai army in organization, pay and efficiency.(17)

THE FEDERAL BUREAU OF NARCOTICS AND THE CIA

It is one thing to find CIA officers and operatives working hand in glove with foreign drug merchants, but quite another to find a long-standing arrangement between the Federal Bureau of Narcotics and U.S. Secret Services in which the concerns of the latter were always paramount. This had important institutional consequences, particularly within the Federal Bureau of Narcotics which provided excellent "cover" for CIA agents.

A case in point is the career of Henry L. Manfredi who had been in the Army's Criminal Investigation Division until October 1951. At that time, Manfredi was officially transferred to the Bureau of Narcotics.(18) But as a letter from CIA headquarters (signed by Richard Helms as Director of **Secret Group One**) to the Secretary of the Treasury in the autumn of 1967 disclosed, Manfredi's 1951 transfer was to the Agency not the Narcotics Bureau.(19) Ironically enough, the found corresspondence is a CIA recommendation to the FBN on Manfredi's behalf:

> I would like to bring to your personal attention the transfer of Mr. Henry L. Manfredi to a career position in Treasury's Bureau of Narcotics after more than fifteen years service in this agency. Mr. Manfredi's contribution to the attainment of the U.S. Government objectives abroad has been outstanding. I am confident that the Bureau is getting a superior officer and although he will be sorely missed by his colleagues here we are pleased that he is assuming an important position for which we feel he is uniquely qualified. I wish also to express my keen appreciation for the fine cooperation the Bureau of Narcotics, most recently under its present director Mr. Henry L. Giordano, has furnished us over the years. This is a particularly significant factor in the success of Mr. Manfredi's mission.

Manfredi was hardly an exception. He was part of a tradition of asymmetrical relations between the FBN and American clandestine organizations which began with the outbreak of World War II in Europe. Much of this hidden history can be glimpsed by looking very briefly at the careers of several significant FBN officers including the flamboyant George Hunter White and his FBN supervisor Garland Williams.(20) Williams and White both played key roles in the organization of American counter-espionage efforts and the training of special agents. As the head of the Corps of Intelligence Police, Williams took that raggle-taggle group and transformed it into the Army's CounterIntelligence Corps. White followed Williams into the wartime clandestine world joining as soon as practicable the Office of Strategic Services (OSS).(21)

Williams designed a tough basic training program which specialized in teaching "special forces" the techniques of silent killing and demolitions. Advanced work called for agents to learn the ways of foreign undercover work including sabotage of enemy activities.(22) When White joined, he was initially assigned to a training facility in Oshawa, Canada, run by British Intelligence. White, in a famous remark, called the place the "Oshawa School of mayhem and murder."(23) He finished this first round of training and soon was chosen to head the New York X-2 training operation. X-2 was the designation given to the OSS counter-espionage group. White's headquarters for OSS X-2 training was the New York office of the FBN. Eventually he became the "director of all OSS counter-espionage training."(24)

To find key FBN officials hip-deep in counter-espionage activities during the war is interesting but hardly extraordinary. But it is remarkable and revealing to find them still diligently at work in the same activities years after. Their ongoing, post-war, clandestine work indicates the subordination of U.S. overseas narcotics operations to anti-Communist endeavors. The FBN hierarchy, from Commissioner Harry Anslinger through Williams, White and many of the men they recruited and trained as narcotics agents, functioned as a counterintelligence unit, attached to the CIA and, at times, the Army. Garland Williams, it appears, went into deep cover after the Korean War in order to help establish the counter-espionage capabilities of the Office of Public Safety located in the Agency for International Development (AID). Commissioner Anslinger, on the other hand, working with former OSS agents among others, set up a private intelligence organization (doubtless in close coordination with the CIA) which was especially concerned with Middle East projects.(25) George White's double life was perhaps the most bizarre as his "diaries" indicate. After his death in 1975, they were given by his widow to an electronics museum on the campus of Foothill College in Los Gatos, California.(26)

These little leather-bound volumes were White's appointment calendars--daily reminders of meetings, tasks, and expenses. The conjunction of his FBN work and Intelligence events is uncanny. In 1948, for instance, while the CIA was conducting one of its first operations, the subversion of Italian politics, White showed up in Rome for a meeting with his former OSS boss William Donovan.(27) More importantly, however, the diaries disclose that White, and his FBN protege, Charles Siragusa, were part of the "inner circle" of CIA officials who planned and carried out various lethal secret operations.

These activities, which in 1953 were known collectively as MK-ULTRA, began in 1950 with Project Bluebird directed by Sheffield Edwards.(28) Its ostensible purpose was to determine if certain drugs might improve interrogation methods. Within the year the direction of Bluebird passed to Allen Dulles who coordinated it with another drug experiment the Agency conducted for counterintelligence purposes named Project Artichoke.

In April, 1953, Dulles, now the head of the CIA, instituted a program for "the covert use of biological and chemical materials" on Americans as part of the Agency's continuing efforts to control behavior. The decision to drug unwitting subjects spawned numerous efforts which were administered by Dr. Sidney Gottlieb who would become chief of the CIA's Technical Services Staff in the 1960s.(29) Gottlieb took control of these new ventures dubbed MK-ULTRA as well as the older Bluebird and Artichoke operations, and yet newer ones called Project Chatter and MKNaomi.

There are several common threads linking these projects together: drug experimentation is one, political assassinations another. As John Ranelagh in his massive study of the Agency noted, the CIA found that "the doctors and biologists in the Technical Services Staff working on MKULTRA subprojects were ambitious to press the frontiers of their disciplines even further, to the point of 'executive action' capability--the agency's in-house euphemism for assasination."(30) It was not only the scientists who wished to move these projects to their logical end; members of the FBN hierarchy were also involved in these affairs, especially White and Siragusa.

There were a number of reasons for White's recruitment into MK-ULTRA, but the most important one was historical. Drug experimentation for counterintelligence purposes had been one of his important OSS missions.(31) Even before the Japanese attack on Pearl Harbor, General Donovan had approved a "truth drug program." Eventually, the plan was put under the supervision of Donovan's "chief scientist," Stanley Lovell and a committe which included the Director of FBI laboratories and FBN Commissioner Anslinger. White was the field officer responsible for testing various subjects, primarily with derivaties of cannabis. He also tested quite a lot of the "T" drug, as it was called, himself.

White's "diaries" show when and where he was contacted by the post-war architects of drug experiments and assassinations. In New York on March 20, 1950, he met with Allen Dulles and the following day had instructions for an FBN agent soon to be sent abroad who also doubled as a CIA operative. Indeed, on January 27, 1951, the FBN established its first overseas office sending Bureau agents Charles Siragusa, Joseph Amato and Martin Perna to Rome. Siragusa was in charge of the undertaking.

In November of 1951 White marked various meetings with the CIA in his calendar. A few months later he appears involved in "vetting" potential CIA operatives likely connected to Artichoke and Bluebird. These kinds of entries continue through April. However, in the middle of that month there are notes indicating his more complete induction into the projects. On April 15, 1952, he met in Boston with Stan Lovell and Colonel Harry Reynolds at what White described as the CIA's Algonquin Club. About three weeks later there was another CIA get together in Boston which was quickly followed by a talk between White and Gottlieb in Hartford, Connecticut. According to his notes, White called Anslinger the day after this meeting to inform him about the CIA drug ventures.

Through June and July, White had a series of meetings dealing with the projects. Late in August, the cast of characters directly involved in these experiments or knowledgeable about them broadened. White travelled to Washington, D.C. for a morning meeting with James Murphy, former head of OSS X-2. He and Murphy were later joined by James Jesus Angleton, dean of CIA counterintelligence. Angleton and White were old and apparently good friends, having met many times during the war and after. White's diaries reveal their friendliness, their many dinners in New York's Chinatown. The particular discussion in Washington that August between these veteran counterintelligence officers concerned "a special teaching assignment" for White. Contacts between White, Murphy and Angleton continued. On Thanksgiving Day (November 27, 1952) White had dinner with Angleton on Long Island. The following day White's diary entry states that after nine in the evening he had taken LSD--"I had a delayed reaction" he wrote.

Meanwhile, it appears Gottlieb had an initial problem in securing White's formal assignment to what would be called the ULTRA projects. Apparently there were some within the CIA less enamored with White than Dulles, Angleton and Gottlieb. No matter. The difficulties were soon resolved and by the summer of 1953 ULTRA and White were moving comfortably forward. White had a new CIA codename, Morgan Hall, and an apartment on the fringe of Manhattan's Greenwich Village. He used this apartment for ULTRA experiments which seemed to consist largely of slipping LSD "mickeys" to unwitting subjects. White celebrated his 45th birthday on June 22, 1953. That day he phoned his old OSS boss, General William Donovan, from the ULTRA apartment to remind him of a meeting they were scheduled to have. The circle of those knowledgeable of the Agency's chemical experiments had obviously included Donovan who seems never to have been in the dark about anything clandestine going on at CIA.

White left New York during the mid 1950s for San Francisco where he continued his Morgan Hall activities from two locations, one described as "a national-security whorehouse on Telegraph Hill."(32) The San Francisco CIA/drug venture was aptly called Operation Midnight Climax, according to commentator Warren Hinckle, and White who was never shy about his activities was characterized as "the best party-dog spy the CIA ever had." Hinckle noted that "White, who had the face of a friendly bulldog, was by day the head of the Federal Narcotics Bureau here. By night, pop! Through the looking glass, he was a Kojak for the CIA, supervising wild drug experiments on unsuspecting johns procured by government-hired prostitutes who subjected their love objects to psychochemicals including LSD aerosols sprays, diarrhea-inducing drops and drug-coated swizzle sticks."(33) When White left

New York, the CIA's east coast experiment did not miss a beat. The ULTRA apartment in the Village was taken over by Charles Siragusa who had been brought back from Rome to supervise the ongoing drug experiments in New York.

White's circle of CIA friends did indeed move forward into the arena of political assassination as Ranelagh states. No one was more eager to kill, it seems, than Gottlieb. He worked on "toxins to poison Fidel Castro and impregnated a handkerchief with poison to kill an Iraqi colonel."(34) Gottlieb also tried to assassinate Patrice Lumumba of the Congo using deadly bacteria. By this time, the CIA had developed several assasination programs. The one involved in the Lumumba plot was named Executive Action later subsumed under the cryptonym ZR/RIFLE. The principal agent in charge of ZR/RIFLE field operations was known as QJ/WIN, supposedly a "foreign citizen with a criminal background recruited by the CIA for certain sensitive programs."(35) Although no one has yet satisfactorily identified QJ/WIN, one thing about this mystery seems certain. Charles Siragusa was either the mysterious QJ/WIN himself as some claim (including a former high official in the FBN and DEA) or at the least controlled him.(36) White's diaries, by the way, indicate that Gottlieb visited him in San Francisco just a short while before and then again right after his African mis-adventure to kill Lumumba. On the second visit Gottlieb was accompanied by Siragusa.

The FBN under the leadership of Anslinger, Williams, White, Siragusa, etc., was one of the lead organizations involved in foreign counterintelligence activities in pursuit of anti-communist objectives. So zealous was the Bureau in these matters that its highest officers committed numerous crimes ranging from drug distribution and drug use to planning foreign assassinations. To know that they committed these crimes while wearing their CIA hats in their Morgan Hall apartments is not comforting. The Narcotics Bureau had been subordinate to the CIA since the Agency's founding. But not everything was cover and counterintelligence. FBN pronouncements about overseas drug producers were often ludicrous propaganda statements about Communist intentions.

The China Card

Drug villians shifted with the tides of foreign policy. That is why early in the Cold War Communist China was alleged to be hard at work undermining American society by dumping drugs here at an alarming rate. This cuckoo theme, published by the Narcotics Bureau (doubtlessly under the CIA's spell), was first tried in the 1950s. For example, in 1954 Anslinger stated,

> The three-fold increase in some areas in the land
> devoted to the cultivation of the opium poppy in
> Communist China, the establishment of new

heroin factories in Communist China, the
continuation and expansion of a 20-year plan to
finance political activities and spread addiction
among free peoples through the sale of heroin
and opium by the Communist regime in China,
and the extension of the same pattern of narcotic
activity to areas coming under the jurisdiction of
Communist China has mushroomed the narcotic
menace from Communist China into a multi-
headed dragon threatening to mutilate and
destroy whole segments of populations from
whom the danger of addiction through ready
availability of drugs had been removed during the
past 40 years by the uncompromising work of
the narcotic enforcement authorities in the free
world.(37)

Thus while CAT was flying Thai opium in Southeast Asia, while
George White and the Agency were doping the unsuspected with LSD, and
while the remnant of Chiang Kai-Shek's Koumintang army which had been
pushed into Southeast Asia positioned itself to control more and more of the
region's opium traffic with American support, FBN press releases gobbled up
by right-wing commentators and politicians called Red China the world's
greatest narcotic menace.

But no matter how outlandish the charge, and how often it was shown
false and intentionally misleading, the truth was not a deterrent. Always
simmering, the tale of Communist perfidy in drugs returned with a rush in
1972 when a dangerous "crackpot" organization, the World Anti-Communist
League, with support from the American clandestine community and its clone
the Korean CIA, once again waved the "bloody Communist drug flag."(38)
Using their Taiwan Chapter as the disseminator of narcotics information, it
distributed bizarre pamphlets entitled "The Chinese Communist Plot to Drug
the World," "Communist China and Drug Traffic," "True Facts of Chinese
Communist Plot to Drug The World," and in 1976 "Chinese Communist
Criminal Acts in Drugging the World."

The "True Facts" pamphlet contains the following pseduo-information
under the headings THESE ARE THE ACCUSERS. In 1950, the U.S.
reported to the U.N. that "Chinese Communists smuggle large amounts of
opium into Burma and other countries." Four years later, Commissioner
Anslinger stated the Red Chinese are "massively exporting opium in an
attempt to drug the world." Almost a decade after that, Harry Giordano
(Anslinger's successor) identified in the pamphlet as the Chairman of the
U.S. Commission on Narcotic Drugs repeated Anslinger's comments.
Giordano was quoted stating "the Chinese Communists are engaged in
massive narcotics trafficking operations."

The litany of bunkum continued until the early 1970s when another problem appeared--President Nixon's rapprochement with China. The World Anti-Communist League was prepared, though, having American spokesmen ready to claim in one dumb remark or another that the Nixon doctrine was a "policy of disguise," pre-formulated to prevent government officials from revealing the Chinese Communists' true role in drug production and smuggling. Right-wing politicos, some on the Taiwan pad others supported by the South Korean CIA, angry with Nixon for dealing with China, issued these goofy statements to an increasingly incredulous world. Much to the despair of the Taiwanese and other Communist haters around the world, mainland China had finally proven acceptable to a broad spectrum of Americans. Improved relations with China stemmed the flow of drug accusations from all but the most cantankerous anti-Communists. In fact, David Musto recently wrote that the U.S. actually defended the Peoples Republic of China from allegations about drug running that had been published in a Soviet newspaper in the early 1970s.(39)

THE CONTEMPORARY LATIN AMERICAN SCENE

That doesn't mean the broader theme--Communist imperialism equals drug smuggling to weaken the West's resolve--dried up, however. Throughout this decade the same tune has been played again and again, and a lot closer to home. Many Latin American countries have been targeted here for special attention in this regard. During 1988, for example, Panama, Mexico, The Bahamas, Jamaica, Paraguay, Bolivia, Peru and Colombia have all been identified as "problem" nations with regard to their contributions to the U.S. drug problem through "narcotrafficking." A key term that has been used repeatedly in press releases regarding these concerns is "narcoguerillas." Apparently the term was introduced in 1984 by Lewis Tambs, U.S. Ambassador to Colombia, who then announced the spectacular success of a raid on a jungle-based drug complex known as Tranquilandia. Tambs claimed that narcoguerillas--communist rebels--had been guarding the facility. The raid on the complex netted some 27,000 pounds of cocaine and was said to be largest coke bust ever.(40) The original Tambs' charge was never proven, yet the term stuck and became a stock phrase in the Reagan administration's rhetoric about the war on drugs being waged in Latin America.

In recent history Colombia has presented perhaps the most significant and lasting problem in terms of Latin-American based U.S. interdiction efforts. One writer has referred to Colombia as the "lynchpin of America's narcopolitics." Colombia has been continually identified as a major contributor to our domestic drug problems by serving as a home and base of operations for many of the world's most notorious cocaine traffickers, presently dominated by the Medellin and Cali cartels. Recent newspaper accounts indicate that the two factions are now involved in a dispute over control of the New York market. In this dispute, gang members have been killing one another and providing enforcement officials with information regarding the other group's drug shipments.(41)

Colombia's reputation for complicity in drug running has been almost legendary in this decade. That nation's current Attorney General, whose predecessor was shot to death allegedly by drug traffickers, suggested earlier this year that perhaps the drug dealers would have to be appeased and that negotiations with them should include discussions of the possible legalization of cocaine. A U.S. State Department official responded, "It's an outrage, an absolute outrage. This man is charged with upholding the law of the land and he's saying that the job's too difficult."(42) Another Colombian government representative quickly denied that the Attorney General's remarks in any way reflected that government's policies or commitment to narcotics enforcement.(43) The current Vargas regime is especially concerned with bad relations with the U.S. and has actually gone so far as to hire a consulting firm in the U.S. to help improve that country's image here regarding their anti-drug efforts.(44)

Linking Colombia's drug lords with communism occurred initially during the bloodbath that took place at the Palace of Justice on November 6-7, 1985. During that confrontation more than 100 people were killed including the President of the Supreme Court, other magistrates and their assistants, canteen workers, government soldiers and guerilla members of the M-19 communist organization. At the time the M-19 soldiers were said to have been sponsored by drug traffickers who were distressed over a pending extradition treaty with the U.S.(45) That charge has been repeated ever since, despite the fact that two commissions of inquiry never found any evidence that M-19 had been working either with or for drug traffickers! With regard to these events, Dr. Rosa Del Olmo states,

> It is worthwhile that what happened in Colombia
> is not unique in this respect. Attempts at protest,
> in Latin America, are all too often smeared--by
> those with   vested interests in suppressing
> criticism--with the labels of drugs and subversion,
> along the lines imputed and/or believed to be set
> by the U.S. government.(46)

The communism--drug theme surfaced again in the 1986 indictment of the notorious Colombian smuggler Carlos Lehder and the Medellin Cartel "which consisted of controlling members of major international cocaine manufacturing and distribution organizations."(47) Charged along with Lehder and other cartel members was Federico Vaughan an "assistant to Tomas Borge, Minister of the Interior of Nicaragua."

Vaughan supposedly helped the cartel establish cocaine conversion laboratories and distribution facilities in Nicaragua. The argument in the indictment claims that actions by the Colombian government moved the cartel to set up part of its shop in Nicaragua. Later in the indictment, however, the following statement about cocaine transshipment and importation into the U.S. is made:

> The Cartel maintained airstrips in South and Central American and made arrangements for transshipments of cocaine and refueling of aircraft and vessels in the Caribbean Islands and in Mexico. Pilots retained by the Cartel would and did fly cocaine from processing laboratories in South and Central America to trans-shipment points in Colombia, the Bahamas [sic], Turks and Caicos Islands, Jamaica and Mexico. At these locations, aircraft would be loaded and unloaded, serviced, and protected by Cartel employees and independent organizations, including service organizations headed by <u>officials of the aforementioned countries</u> (my emphasis).(48)

The Reagan administration worked hard to implicate Nicaragua in the Medellin cartel's activities. This was done partly to counter the growing evidence of Contra drug smuggling, and to convince the wary that the Sandinista government was capable of anything in its war with America. It now appears likely that much of the Vaughan evidence was fabricated. In addition, the only allegedly credible witness against him was Adler Berriman Seal who was murdered in Louisiana by cartel gunmen. But even if Vaughan was guilty, his participation pales next to that of the "officials of aforementioned countries." Where are the Jamaican, Mexican, and especially Bahamian officials? The answer is obvious. They are missing because the administration has politicized the Caribbean drug traffic issue wildly overstating Communist interest and minimizing the clear responsibility of important regional friends. The administration's intent was announced in the 1984 Republican platform which identified the "international drug traffickers who seem to irritate the Reagan Administration the most"--Cuba, Bulgaria, the Soviet Union, and Nicaragua.(49)

It is well known that Florida prosecutors have been anxious to indict the Prime Minister of The Bahamas, Lynden Pindling, for quite some time. Pindling's partnerships with drug smugglers, in particular Carlos Lehder, have been alleged for years. The House Select Committee on Narcotics has debated the Pindling issue "in camera" and instead of holding the public hearing which would doubtlessly establish Pindling's participation and enrichment in one drug conspiracy after another, have chosen another path.(50) For reasons that are not clear, the Committee lent its most knowledgeable drug consultant, John Cusack, to the Bahamian government.

The Bahamian story reveals the limits of narcotic enforcement as it clashes with the most fundamental element in America's Caribbean policy-- regional anti-communism. Pindling has likely been protected because of The Bahamas long-standing role in the U.S. struggle against Cuba. It is no secret that The Bahamas has provided a base for American sponsored infiltration and sabotage teams bound for Cuba.(51) The Bahamas strategic importance goes beyond this, however. The U.S. has long had a vital submarine listening post on Andros Island, as well as other intelligence installations in The Bahamas. These stations have given Pindling significant leverage with both the State and Defense Departments which has been used to urge restraint on the implementation of U.S. drug policy.

Colombia, The Bahamas, Nicaragua, and so on provide only some of the many contemporary examples of U.S. drug policy confusion. Bolivia is another. In that nation which contains some of the most important growing regions for coca, similar events have unfolded. There former Air Force Major Clarence Merwin was hired in 1984 by the State Department's Bureau of International Narcotics Matters to organize, train and supervise an elite paramilitary force to conduct raids on the various growers and refiners of coca.(52) The Coca Reduction Directorate had been created in the Bolivian government as a result of four narcotic treaties signed with the U.S. on August 11, 1983. It called for a five-year program for reduction in coca production to a level that would only support domestic chewers of the leaves (it was estimated at the beginning of this process that local production was then more than twenty times that required for indigenous consumers).(53)

Major Merwin trained the Mobile Rural Patrol Unit, popularly known as the Leopards, to serve as a model in the creation of similar anti-drug units elsewhere. He found every attempt made to utilize this force effectively thwarted. Not only was there corruption within its ranks (many officers had accepted bribes from drug barons), but the Bolivian government repeatedly failed to live up to its end of the bargain in providing supplies and paying the troops. According to Merwin, the U.S. government representative, Ambassador Edwin Corr, spent his time trying to "maintain stable relations with unstable Bolivian governments" and refused to force them to honor the original commitment.(54)

The raids conducted by Merwin's Leopards failed to net any significant drug pushers and concentrated instead on peasants and low level producer-seller-processors. Based on these experiences, Merwin stated "there are simply no sanctions being applied against any but the lowest level of traffickers."(55) He quit his position in 1986 as it had become clear to him that he was not accomplishing anything of significance in Bolivia and that the U.S. government representatives were not supporting his efforts. At the end of his service the estimated number of acres of coca under cultivation in Bolivia, far from being dramatically restricted, had actually increased by at least 10 percent over the pre-Leopard period. Since Merwin's departure the Leopards were involved in Operation Blast Furnance, a much publicized series of raids that was said by the State Department to have "disrupted the local cocaine industry." Actually only empty labs were found and not one important trafficker was arrested during the 256 separate raids.(56)

The lesson in Bolivia, like The Bahamas, is that relations with apparently friendly anti-communist governments will never be sacrificied for drug control. Despite the obvious unwillingness of the present Bolivian government to rigorously pursue local traffickers, and at times the direct involvement of top government officials in the cocaine trade (as occurred in the famous cocaine coups of 1981), U.S. officials have not pushed the issue. Clearly something more important is on their agendas; we believe that traditional anti-communism fits the bill, although carried out under a new guise.

The contemporary examples discussed here are instructive in putting together the puzzle of U.S. drug policy in the region. The Reagan White House has been effective in linking in the public mind drug trafficking and communism. The charges that communist guerillas or regimes such as in Nicaragua are either directly involved in the drug trade or at least protecting drug producers in exchange for sorely needed currency to support their cause or government are repeated often enough to have assumed a life of their own.

It is obvious why the administration wants to present this linkage of very long standing. Foreign drug enforcement carried out by U.S. paramilitary units, directing and training indigenous forces, is the **cover** for hemispheric counter-insurgency efforts. High technology weapons and advanced radar equipment, Delta force teams, the stationing of American naval, customs and coast guard ships ever closer to key South American ports, are in place to counter the Communist menace. The effectiveness of the administration in convincing others that communism and narcotics march together may determine the success of the entire operation.

* * * * * * * * * * * * * * * * * * *

NOTES

1. David Kline, "How to Lose the Coke War," Atlantic Monthly (May 1987), 23.

2. This theme has been explored to one degree or another by other writers as well. See Johnathan Marshall, "Drugs and United States Foreign Policy," in Ronald Hamowy (ed.) Dealing With Drugs (D.C. Heath and Co., 1987), 137-176.

3. Bruce Michael Bagley, "Reflections on the U.S. Latin American Drug Trade: Losing the War on Drugs," in Miami Report II: New Perspectives on Debt, Trade, and Investment by the North South Center Graduate School of International Studies, University of Miami, Coral Gables, Fla., May 1988, 115.

4. See Alfred W. McCoy with Cathleen B. Read and Leonard P. Adams II, The Politics of Heroin in Southeast Asia (Harper & Row, 1972).

5. McCoy; Frances Fitzgerald, Fire in the Lake (Random House, 1972); William Corson, The Betrayal (W. W. Norton, 1968).

6. Author's interview with William Corson in Washington D.C., 1982.

7. See the William Donovan Papers, Carlisle War College, Carlisle, Pa., Paul L. E. Helliwell to Bernard B. Fall, "Letter," October 14, 1954, in which it is recounted that Helliwell was awarded the Oak Leaf Cluster to the Legion of Merit, the Asiatic Campaign Medal with two bronze stars, and similar awards for outstanding intelligence work in Egypt.

8. See William R. Corson, Armies of Ignorance: The Rise of the American Intelligence Empire (Dial Press, 1977), 221-290.

9. Martindale & Hubbell, Law Directory Florida (1949), 453.

10. William L. Leary, Perilous Missions: Civil Air Transport and CIA Covert Operations in Asia (University of Alabama Press, 1984), 3.

11. Leary, 67-68.

12. Bob Woodward and Scott Armstrong, The Brethern: Inside the Supreme Court (Avon Books, 1981), 88.

13. Leary, 72.

14. Leary, 82.

15. Leary, 129; and see Martindale & Hubbell (1952), 661.

16. Martindale & Hubbell (1952).

17. Noam Chomsky and Edward S. Herman, The Washington Connection and Third World Fascism (South End Press, 1979), 220-222.

18. This transfer "in accordance with Department of Army message #28367," was signed by Marvin A. Ruckman, Assistant Civilian Personnel Officer.

19. The Helms correspondence dealing with Manfredi's CIA service and FBN cover was left, inadvertently no doubt, in a box of otherwise innocuous material located at DEA headquarters in Washington, D. C. I recorded the correspondence and transcriped it in the Spring of 1988.

20. On White see John McWilliams and Alan A. Block, "All the Commissioner's Men: The Bureau of Narcotics and the Dewey-Luciano Affair, 1947-1954," Intelligence and National Security (April 1990); and Alan A. Block and John McWilliams, "On the Origins of American Counterintelligence: Building a Clandestine Network," Journal of Policy History (1989).

21. Block and McWilliams, "On the Origins of American Counterintelligence."

22. Corson, 81.

23. This famous remark found in the George White Diaries (1943) located in the Perham Electronics Museum, Los Gatos, California.

24. David Stafford, Camp X: Oss, "Intrepid", and the Allies North American Training Camp for Secret Agents, 1941-1945 (Dodd, Mead and Co., 1986), 82.

25. See Block and McWilliams, "On the Origins of American Counterintelligence,"; and on the Agency for International Development see A. J. Langguth, Hidden Terrors (Pantheon Books, 1978).

26. I reviewed these volumes on two occasions in 1986 and 1987 and tape recorded summaries of all relevant material. Transcriptions of these recordings were then made and have been used in the following analysis.

27. The operation aimed to crush the power of the Italian Communist Party and elevate the Christian Democratic Party.

28. See John Ranelagh, The Agency: The Rise and Decline of the CIA (Simon and Schuster, 1986), 204.

29. Ranelagh, 202-216; also see Martin A. Lee and Bruce Shlain, Acid Dreams: The CIA, LSD and the Sixties Rebellion (Grove Press, 1985).

30. Ranelagh, 207.

31. McWilliams and Block, "All the Commissioner's Men."

32. Warren Hinckle, San Francisco Examiner (7 November 1985).

33. Hinckle.

34. Ranelagh, 211.

35. U.S. Senate Select Committee to Study Governmental Operations with Respect to Intelligence Activities, <u>Alleged Assassination Plots Involving Foreign Leaders: An Interim Report</u> (Government Printing Office, 1975), 189.

36. Interview with John T. Cusack former FBN and DEA officer currently on leave from the House Select Committee on Narcotics.

37. The Anslinger quote can be found in Richard L-G. Deverall, <u>Red China's Dirty Drug War</u> (Tokyo, 1954).

38. For an informative history of the League see Scott Anderson and Jon Lee Anderson, <u>Inside the League</u> (Dodd, Mead and Co., 1986).

39. David Musto, "The History of Legislative Control Over Opium, Cocaine, and Their Derivatives," in Hamowy, 68.

40. Merrill Collett, "The Myth of the Narco-Guerillas," <u>The Nation</u> (13/20 August 1988), 129.

41. Alan Riding, "Gangs in Colombia Feud Over Cocaine," <u>The New York Times</u> (23 August 1988), A-1. For an interesting discussion of historical and contemporary violence in Colombia, see E. J. Hobsbawm, "Murderous Colombia," <u>The New York Review of Books</u> (20 November 1986), 27-35.

42. Elaine Sciolino, "Colombian Official Talks of Legalizing Cocaine," <u>The New York Times</u> (25 February 1988), A-11.

43. Sciolino.

44. James Petras, "Colombia: Neglected Dimensions of Violence," <u>Contemporary Crises: Law, Crime and Social Policy</u> 12:3 (1988).

45. Rosa Del Olmo, "The Attack on the Supreme Court of Colombia: A Case Study of Guerilla and Government Violence," <u>Violence, Aggression and Terrorism</u> 2:1 (1988), 57-84.

46. Del Olmo.

47. U.S. District Court, Southern District of Florida, <u>United States of America V. Jorge Ochoa-Vasquez et. al.</u>, Indictment No. 86-697-Cr-Scott.

48. U.S. District Court, 32.

49. "The Communist Connection," <u>The New York Times</u> (13 September 1984), A-17.

50. Interview with staff members of the House Select Committee on Narcotics, 1988.

51. <u>Report of the Royal Commission Appointed on the Recommendation of the Bahamas Government to Review the Hawksbill Creek Agreement</u>, Volume I, paragraph 133, p. 44.

52. David Kline, 22.

53. Kline.

54. Kline, 24.

55. Kline, 27.

56. Kline.

PART VI:
REFLECTIONS ON HISTORY
AND CRIMINOLOGY

Chapter 11
American Criminals Abroad

If a researcher working in the 1960s had asked about the nature of European organized crime the likely answer would have been that organized criminality existed but only as a pale imitation of the American version. This was particularly so when speaking about the United Kingdom, the Scandinavian countries, The Netherlands, West Germany, Belgium, and France (with the exception of the Marseilles region, naturally). Professional criminal gangs in those nations, it was held, were neither prone to violence nor linked to political figures and criminal justice officers as were American racketeers. The American notion of organized crime as a social system composed of professional criminals, politicians and criminal justice officers, and clients anxious for illicit goods and services was thus foreign.(1)

Over the course of the last decade or so, the belief in European civil probity has diminished as organized crime syndicates similar to American ones have been discovered in northern and western Europe by journalists and academic researchers.(2) Criminal syndicates based in cities such as Paris, Madrid, Munich, Hamburg, Brussels, and Milan have been identified.(3) They not only work locally but are often in league with organized criminals from other cities and nations. This should not be unexpected particularly when considering that most mature crime syndicates deal in illicit, stolen or counterfeit commodities like drugs, guns, gems, currencies, and other financial instruments. Such enterprises are not feasible without international links among professional criminals.

And, as ought to be expected, the connections between crime syndicates are dynamic. They closely mimick, in fact, those between firms and individuals engaged in legitimate international trade and finance. Organized crime on the international level is carried out in a constant swirl brought about by the needs of secrecy and security, and the demands of changing markets. Professional criminals at this level are "frequent flyers," primed to shift their activities, open new markets, and secure safe havens. An important example of this restless shifting and moving which took place in the late 1960s and 1970s brought certain American organized criminals to Holland. This surely suggests a vital Dutch underworld already in place and, as we will see, linked to America's most creative criminals.

228

This international linkage is examined by considering first who the American criminals were, and second what they were doing in Holland. Unfortunately, the latter question is only briefly addressed as the Dutch data are slim and speculative. Notwithstanding that, let us look at who the Americans were and what were their criminal enterprises.

THE LANSKY SYNDICATE AND ASSOCIATES

The American organized criminals who incorporated Amsterdam into their international activities were led by Meyer Lansky one of this century's most interesting and innovative gangsters. A brief look at his biography reveals his inventiveness and associates. Born in Grodno, Poland, in 1902, his family migrated to America during the time of increasing troubles for East European Jews. His criminal career started on New York's Lower East Side where he worked as a gunman and thief. It was Prohibition (America's experiment with alcoholic abstinence) which provided Lansky with an opportunity for rapid criminal advancement.

One of his particular strengths was an ability to construct criminal cartels bringing together racketeers normally competitive. He worked with Abner "Longie" Zwillman from Newark, New Jersey, whose bootlegging connections ran to the Bronfman family in Canada owners of Seagram Distilleries. He also teamed with Lewis Rosensteil (who despised the Bronfmans) and several other bootleggers from Boston and Chicago to form another alcohol "consortium."(4) Rosensteil was the founder of Schenleys, an exceedingly large and politically powerful liquor business in the decades after Prohibition. Through Rosensteil, Lansky maintained a substantial financial interest in the legitimate liquor industry. He was joined in this by racketeers from across the country who owned or managed Schenley warehouses in their territories.

Lansky represents the rise of a new generation of organized criminals in the United States; young men who rode to the top of organized crime through bootlegging. Lansky, for instance, constructed his alcohol syndicates well before his 30th birthday. At the same time he did not neglect the organization of gambling his other illicit mainstay. In 1936 Lansky and Zwillman along with Jack Dempsey the former heavyweight boxing champ formed a partnership. They bought a Miami Beach hotel to serve as a front for an illegal casino. One year later, Lansky was running a casino at the famous Hotel Nacional in Havana, Cuba. The U.S. Department of Justice kept a watch on Lansky and reported in 1939 that his Cuban activities included "operating the race track and National Gambling Casino during the Winter of 1938-39."(5) Crime reporter Hank Messick added some detail to this early federal report noting Lansky leased the local racetrack from its real and surprising owner--the National City Bank of New York.(6)

For the diminutive Lansky (he was only 5' 4" tall and weighed no more than 140 pounds) the Depression decade was one of imperial expansion. His illicit businesses migrated from New York to Newport, Kentucky, Southern California, Nevada, Florida, and, as stated, Cuba. His partners and allies were the leading organized criminals from New York, Boston, Philadelphia, Cleveland, Detroit and Minneapolis. Working with him in California and Nevada was one of his oldest friends, Benjamin "Bugsy" Siegel. It was Siegel (murdered in 1947) who dreamed up modern Las Vegas building the first hotel and casino (the Flamingo) on the "strip."

On the eve of World War II, Lansky's influence reached into the world of military intelligence. During the war he worked with a Naval Commander from the Office of Naval Intelligence who was in charge of covert activities in the New York port area. Lansky helped construct an Intelligence/Organized Crime project. He brought several of New York's most infamous racketeers including Joe Adonis, Frank Costello, Willie Moretti, and the Mangano brothers named by Italian authorities as the leaders of an American Mafia into this project.(7)

Five years after the war, the U.S. Congress finally got around to investigating organized crime. A special committee was formed in 1950 under the leadership of Senator Estes Kefauver, a Tennessee Democrat, to monitor organized crime activities across the U.S. The Kefauver Committee's investigators discovered evidence of Lansky's enterprises in several states. They established that Lansky and his brother substantially controlled gambling in Florida's Broward County owning illegal casinos named the Colonial Inn, Club Greenacres, and Club Boheme. The Lanskys and their confederates were also owners and managers of many Miami Beach hotels each featuring a combination of gambling operations--bookmaking, high stakes card games, etc. South Florida was a gambler's heaven.

Joining the Lansky brothers in some of their gambling ventures were several racketeers including the renowned bookmaker Frank A. Erickson. One of the most important gambling entrepreneurs in the nation, Erickson accepted wagers from every major race track in the U.S., as well as bets on boxing, hockey, and baseball.(8) Erickson was a dynamic member of the New York gambling fraternity, a bookmaker's bookmaker handling layoff bets from other gamblers fearful of potential losses.(9)

Another important Lansky partner was Harry Stromberg called at one time Philadelphia's most powerful racketeer. That is how the Assistant Superintendent of the Philadelphia police described him to the Kefauver Committee. He was "the kingpin of gangsters, bootleggers, and rum runners" during Prohibition.(10) The Lansky/Rosen partnership affected many activities in New York, Philadelphia, Miami Beach, Las Vegas and Mexico. For instance, the Federal Bureau of Investigation reported Lansky and Rosen likely financed the expansion of Mexico's heroin industry. The Mexican heroin connection running to New York and Philadelphia started around 1939 and became increasingly important as older drug routes were disrupted by

Nassau. It was managed by Eddie Cellini who was Dino's brother. Eddie, and still another Cellini brother named Goffredo, had had long experience as Lansky employees. Together they had run one of Lansky's premier Cuban casinos.(18) The Nassau casino venture was a small one. It was, however, merely preparatory for the largest casino enterprise in The Bahamas. This was Resorts International a casino and hotel built on a small island which formed the shelter for Nassau's harbor. The gambling license from the Nassau casino was transferred to this new undertaking. Also transferred was Eddie Cellini as casino manager. Although Eddie's work for Lansky was almost as well known as Dino's, he was able to stay in The Bahamas for many years working for Resorts International. The government did pressure the company to get rid of him, but Resorts was exceedingly difficult to budge when it came to Eddie.

The 1964 embarassment had put Dino Cellini on the spot as the best known criminal operative working the Monte Carlo. Dino had to go to appease wounded Bahamian sensibilities. Although he was forced to leave The Bahamas, he was not fired. Dino was sent to London to open a croupier school. The plan called for training a European staff of dealers for the Monte Carlo. This London sojourn didn't last long, however, for British authorities were quite leery of looking ridiculous. How could London tolerate a criminal bounced out of The Bahamas? Thus, Dino was soon deported as an undesirable alien.

This London interlude, however brief, was important for the Holland connection. Joining Dino at the croupier school was Fred Ayoub who had an extensive gambling background in Las Vegas. In time Cellini, Ayoub and other American syndicate criminals linked together with and by Lansky were in Amsterdam. The most significant other was Joseph Francis Nesline the leader of organized crime in Washington, D.C.

Nesline was born in Washington in 1913 and became a bootlegger during Prohibition. Following Repeal, Nesline who was a gambling manager like Lansky, opened illegal casinos in the Washington area. He also killed one of his rivals in the winter of 1951 during a dispute in a gambling club in the Northwest section of Washington. The homicide helped establish Nesline's reputation in the District.(19) From then on he was Washington's most menacing organized criminal.

Nesline's closest associate was a New Jersey born gangster named Charles Tourine.(20) A formidable organized crime figure, Tourine was part of a gang of New Jersey racketeers led by Ruggiero Boiardo which was affiliated with the New York Genovese crime syndicate. Tourine was thus a member of a powerful so-called "Mafia" syndicate while working at the same time with organizations and businesses run by Nesline and Lansky. Gambling was one of Tourine's specialties and he too operated casinos in Havana in the pre-Castro days. Tourine was also believed to be a mob killer. As early as

1945 the New Jersey State Police suspected he was involved in gang assassinations. And fifteen years later, Tourine was the primary suspect in the slaying of racketeer Frank "Buster" Wortman from St. Louis. The murder resulted from a vending machine "war" between Wortman's gang and one from New York.(21) Tourine, like Nesline and others associated with Lansky, found opportunity everywhere. In 1976 he and several former federal and state officials from Alaska were charged with conspiring to establish organized prostitution and gambling for Alaska's pipeline workers.(22)

There is no doubt that Lansky, Nesline, Tourine, etc. made up a very powerful criminal cabal with associations seemingly everywhere. Their notoriety was such that during the early 1960s law enforcement agencies closely monitored their activities. For example, in 1962 narcotic officers following them noted that Nesline's wife and Tourine left the U.S. and travelled to Paris. Later that year, Nesline and Lansky were spotted in England and that summer Nesline and Tourine were observed in Bermuda.

American narcotic officials enlisted the help of their foreign counterparts and the Central Office for the Repression of the Illicit Narcotics Traffic (a division of the French surete) watched Lansky and his wife during a ten day trip they made to Paris in the fall of 1962.(23) American narcotic agents resident abroad also followed Lansky whenever he travelled overseas. In 1965 Agent Albert Garofalo working out of Marseille, Europe's most infamous heroin center, reported on Lansky associates in Nice. Another agent conducted a review of Lansky's gambling associates from Miami Beach, Las Vegas, and The Bahamas. The Federal Bureau of Narcotics suspected they travelled the world on behalf of Lansky in order to buy drugs and arrange smuggling.(24)

AMSTERDAM AND HAMBURG

The links between Lansky and his confederates to European underworlds long preceeded the time they were finally noticed in Amsterdam. Many of these individuals like Zwillman, Rosen, Meltzer, and others not mentioned from New York, Cleveland, Chicago, and Detroit were international organized criminals for decades and had extensive European contacts from the late 1920s on. This was interrupted by World War II and the Nazi Holocaust which in its madness destroyed everything Jewish throughout Europe including organized crime syndicates. It was these connections, far more important for drug smuggling and many other types of organized crimes than those between Italian/American and Italian criminals, which formed the basis and background for Lansky's international criminality.(25)

Lansky and his partners Nesline and Tourine, and employees like the Cellinis and Ayoub exerted an influence in the Amsterdam underworld during the 1970s. Ayoub was issued a four-year residence permit by the Aliens Department of the Amsterdam Police, according to the Interpol office in The Hague, on 2 October 1974. He was joined there by his second wife, a British subject, and his son Peter who was also a professional gambler. They too had residence permits.(26) Interestingly enough, however, official attention seemed not to be aroused until 1978. In that year there was considerable notice taken as the cable traffic between Interpol offices in The Netherlands, Washington, Wiesbaden Germany, Belgrade Yugoslavia, and Paris attest.

It was reported, for example, in the spring of 1978 that Dino Cellini had been staying in Amsterdam for some time but had recently left for the U.S. because of ill health. He died soon after. His brother Goffredo and Joe Nesline remained behind in residence at the Caballa Club, 98 Oudezijds Voorburgwal. It was also believed that Fred Ayoub lived at the Caballa Club, though this wasn't accurate. Ayoub and his family were living on a canalboat named Ria moored at Hugo de GrootKade 76.

The Caballa Club was reportedly an illegal casino. It was also either a subsidiary of or in some other manner closely connected to a sex club named the Casa Rosso. As even the most casual visitor to Amsterdam's red light district could have verified, Casa Rosso was the name of a number of sex clubs. Indeed, it would not be surprising to learn that Casa Rosso controlled, at that time, a large segment of the district's sex businesses. Interestingly enough, among the reports claiming Fred Ayoub supervised the gambling at Club Caballa, one pointed out that his canalboat's telephone number was the same as the Casa Rosso's unlisted one.

The proprietor of Club Caballa was Mauritz De Vries born in Utrecht, The Netherlands, in 1935. The Interpol office in The Hague stated de Vries was "very vague about the activities of [Fred, Goffredo and Nesline] . . . . Sometimes he calls them supervisors, another time he says that they come to his club as gamblers." The report added "there are good reasons to assume that they do engage in the running of the Caballa Club."(27) This, it should be added, was categorically denied by Nesline to American police in an interview in the spring of 1979. Nesline claimed no "financial interests in the Caballa Club" which he described as a "legal gambling club located in the red light district of Amsterdam. The club has slot machines, crap tables and bingo, is 3 or 4 stories high, employs 40-50 people and is not open to the public, membership is required."(28) Notwithstanding this denial, the American cabal in Amsterdam was augmented by the presence in 1979 of Dino's other brother Eddie fresh from Caribbean and Colombian adventures.

The group's activities in America continued, of course, unabated. During this same period, for instance, Fred Ayoub's son became the director of the New York School of Gambling located in Manhattan which trained dealers for the new Atlantic City casinos. Robert Ayoub sought to expand this undertaking in late 1977 or early 1978 applying for permission to open

a branch in New Jersey for members of the New Jersey Gaming Bureau.(29)
Truly there was much he could teach them.

While these U.S. gambling projects were moving along, and at the
same time that the Amsterdam activities were proceeding, there were other
European connections which became known. Through a trace on Nesline's
phone in Bethesda, Maryland, the police discovered the Amsterdam
enterprises were part of wider European ones. In particular, the syndicate
was in business with organized criminals operating in Hamburg, West
Germany. The primary contact was Wilfried Schulz "owner of restaurants in
the Hamburg amusement centre" (so Interpol Wiesbaden delicately described
Hamburg's red light district) and a fight manager. Schulz was believed to be
a thief, pimp, gambler, and extortionist. At some point in 1977 Schulz was
held in pre-trial confinement for tax evasion.

The Hamburg organized crime figures in contact with Nesline and thus
Tourine, etc., included Daviad Dargahi who was identified by Interpol
Teheran as Davoud Dargahi born in Teheran in 1932. Dargahi had been
living in Hamburg since 1957 and was associated with Schulz in organizing
boxing matches and various criminal activities. Both were described as having
"considerable influence on the Hamburg criminal milieu and the gambling
hall scene."(30) Moreover, Iranian authorities claimed Dargahi was involved
in the smuggling of heroin to the U.S. They also reported he was closely
associated with the underworld in Milan, Italy.(31) Dargahi's Italian
connections, according to the German police, included Giuseppe De Giorgio
an older Italian originally from Naples "said to have relations to the so-called
Mafia in the USA."(32)

The illicit interests of this group of American criminals we have been
charting in Amsterdam and now Hamburg encompassed gambling, narcotics
trafficking, pornography, prostitution, and the fight game. Even this last point
opens yet another series of illicit connections. Consider that Nesline and
Tourine worked with Nigerian fighter and former world middleweight
champion Dick Tiger in a casino in Nigeria, and that Nesline bought an
interest in light-heavyweight champion Bob Foster from the fighter's manager
Morris Salow a known bookmaker and loanshark.(33)

FINAL THOUGHTS

One might well wonder about the significance of American criminals
found working in Holland and linked to Hamburg, Milan, etc. Without
wishing to stretch the data too far, it seems that these relationships suggest
the following: European organized crime more closely resembled that found
in America until the Nazi interregnum. European organized crime syndicates
were smashed along with much else in European society, but only for a short
while. There were of course significant changes in the ethnic makeup of

syndicates before and after the war. But with it all, the Lansky, Nesline, Tourine, Cellini, Ayoub connections in Amsterdam more closely resembled the way things were before the war and probably were again after a decade or two.

This discussion began by noting that it was long believed that parts of Europe had escaped the kind of organized crime found in the United States. Post-war American criminology which was rigidly ahistorical suggested this, and it was then echoed back by European criminologists who seem to have borrowed far too much from overseas. Ahistoricism is one cause of this narrow view; there are others. Perhaps the most significant other is **access to data.**

American municipal government along with city police have always been venal. We know this because there is an American tradition of investigative reporting that goes back to the turn of the century. This penchant for viewing political and economic institutions with a jaundiced eye, of assuming corruption unless proven wrong, is an interpretation of reality. It is an interpretation constantly reinforced, ironically enough, by the government itself which feeds incriminating information at a remarkable clip to reporters and researchers. American governments are both corrupt and open. Nowhere else on earth is it as easy to get access to secret and confidential documents than in the U.S. and to publish them. This naturally allows confirmation of the worst suspicions; there is an endless revelation of crookedness. It would seem that Americans believe secrecy a greater evil than criminality.

There is no doubt that American government proceeds on the basis of leaked information as the distinguished reporter Harrison E. Salisbury persuasively stated.(34) Writing about the publication of the Pentagon Papers and the titanic court case which ensued, Salisbury noted the following:

> No one, no judge, no ordinary citizen, no official, no editor after reading [New York Times editor Max] Frankel's case-by-case analysis of how government really operated could longer hold the delusion that the classification system was designed to keep secrets from the press. They could not but accept Frankel's common-sensical view that the principal purpose of "classification" was to create a stock of "secrets" which could be traded to reporters for headlines .... Frankel wrote: "Without the use of 'secrets' there could be no adequate diplomatic, military and political reporting of the kind our people take for granted . . . there could be

> no mature system of communication between the
> Government and the people.(35)

This style of governance leads me to wonder how much the distinctions drawn between the American polity and European ones have been affected by the availability of incriminating information. The comparative question about organized crime thus rounds itself into a variant of the classic "dark figure" question in criminology. Are the distinctions seen based on comparable reliable information or not? I think the likely answer is not.

* * * * * * * * * * * * * * * * * * *

NOTES

1. For an explanation of the paucity of research on European organized crime through the 1960s and mid 1970s at least see Alan A. Block and William J. Chambliss, Organizing Crime (Elsevier, 1981) especially 117-134. On the comparison between U.S. and European organized crime see Joseph Albini, "Mafia as Method: A Comparison Between Great Britain and the U.S.A. Regarding the Existence and Structure of Types of Organized Crime," International Journal of Criminology and Penology (1975); and John H. Mack with H. Kerner, The Crime Industry (Saxon House, 1975).

2. For an early discussion on organized crime in Sweden see Block and Chambliss, 135-159; also see M. Clarke, "Syndicated Crime in Britain," Contemporary Crises (1979); and most recently the paper by Cyrille Fijnaut, "Organized Crime: A Comparison Between the United States of America and Western Europe," (presented at the International Society of Criminology meeting, Hamburg, West Germany, 1988) summarizes many of the contemporary themes on European organized crime.

There are historians writing about the international traffic in narcotics during the first four decades of the twentieth century whose work often has much to say about European crime syndicates. This is an emerging literature to say the very least; see particularly Terry M. Parssinen and Kathryn S. Meyer, "International Narcotics Trafficking in the Early Twentieth Century: Development of an Illicit Industry," Past and Present (fortcoming). Although not historians, one should also consider the fine work by Anthony Henman, Roger Lewis, and Tim Maylon, Big Deal: The Politics of the Illicit Drug Business (Pluto Press, 1985).

3. See Maurice Punch, Conduct Unbecoming: The Social Construction of Police Deviance and Control (Tavistock) and Fijnaut note 2. Additional evidence of criminal syndicates active in the cities mentioned can be teased

from material in C.J.C.F. Fijnaut and R.H. Hermans (eds.) <u>Police Cooperation in Europe: Lectures at the International Symposium on Surveillance</u> (Van den Brink, 1987) paying particular attention to statements dealing with "major organised, cross-border criminality" (p. 14), the recognition that European police need to coordinate activities in order to curb "major organised crime, with its international branches" (p. 17), and to Fijnaut's essay in this collection, "The Internationalization of Criminal Investigation in Western Europe," 32-56.

4. New York State Joint Legislative Committee on Crime, "Inquiry into Schenley's Activities," Testimony of Senate Subcommittee Investigator James P. Kelly, February 1971, 7.

5. U.S. Department of Justice, Federal Bureau of Investigation, "The Furdress Case," File no. I. C. 60-1501, 7 November 1939, 25.

6. Hank Messick, <u>Lansky</u> (G.P. Putnam's sons, 1971), 89.

7. See Alan A. Block, "A Modern Marriage of Convenience: A Collaboration Between Organized Crime and U.S. Intelligence," in Robert J. Kelly (ed.) <u>Organized Crime: A Global Perspective</u> (Rowman & Littlefield, 1986).

8. New York Commissioner of Investigation William B. Herland to Mayor Fiorello LaGuardia, "Communication, Re: Frank A. Erickson," 4 May 1939, 4.

9. New York Commissioner, 8.

10. U.S. Senate, Special Committee to Investigate Organized Crime in Interstate Commerce, <u>The Kefauver Committee Report on Organized Crime</u> (Government Printing Office, 1951), 743-44.

11. California State Organized Crime Control Commission, <u>First Report</u> (Sacramento, California, May 1978), 66.

12. U.S. Senate, Permanent Subcommittee on Investigations, <u>Hearings: Organized Crime, Stolen Securities</u> Part III, 20, 21, 22, 27, 28 July 1971, 705-06, 713-19.

13. U.S. Senate, Permanent Subcommittee on Investigations, "Organized Crime Investigations," Testimony of Sergeant John Waymire, Dade County Department of Public Safety, Organized Crime Bureau, 2 August 1978, 317.

14. See Pino Arlarcchi, <u>Mafia Business: The Mafia Ethic & the Spirit of Capitalism</u> (Verso, 1986), 3-19.

15. Florida Department of Law Enforcement and Dade County Public Safety Department, "Supplemental Info, Re: Lansky, Meyer," 1 September 1970.

16. Florida Department of Law Enforcement.

17. Quoted in Hank Messick, Syndicate Abroad (McGraw Hill), 155.

18. Central Intelligence Agency, "Letter from John Edgar Hoover, Director of the FBI to Director CIA, Re: INTERNAL SECURITY--CUBA," 18 January 1961.

19. Washington, D.C., Metropolitan Police Department, "Statement of Facts in Case of Prisoner Joseph Francis Nesline," taken by Detective R. G. Kirby, D.C.P.D. 74798.

20. Federal Bureau of Narcotics, "Examination of Passport Office File of Charles Tourine, Sr.," 6 December 1861.

21. N.Y. District Attorney's Office Squad, "Letter Re: CHARLIE TOURINE AND HIS ASSOCIATES," 30 November 1961.

22. "Ex-U.S. Attorney and 8 Others Indicted in Alaska Prostitution," New York Times 9 August 1976.

23. Federal Bureau of Narcotics, "Translation of Report in Re: Meyer Lansky," Letter #1395, 11 December 1962.

24. Federal Bureau of Narcotics, District No. 5, "File Title--Meyer Lansky," 4 November 1965.

25. On Meyer Lansky see D. Eisenberg, U. Dan, and E. Landau, Meyer Lansky: Mogul of the Mob (Paddington Press, 1976), and Hank Messick, Lansky (G.P. Putnam's Sons, 1971).

26. Direction de la Police Bureau Interpol, 47 Raamweg, La Haye to Interpol WASHINGTON, "Reference Your Letter No. 7522/LS of 5 April 1978 concerning Joseph Francis Nesline and others, Mensaje Postal Condensado, No. 7.323/3465 D/PR, 8 May 1978.

27. Interpol The Hague to Interpol Washington, coded radio message "Nr. 1761 260 20/1501 CMT, 20 March 1978.

28. Montgomery County Department of Police, Office of Criminal Intelligence/Organized Crime, "Continuing Information Report, Subject: Joseph Francis Nesline, File No. OCR 75-12," 30 March 1979.

29. State of New Jersey, Department of Law and Public Safety, Division of the State Police, "Letter from Captain Justin J. Dintino, Re: Robert Ayoub," 28 February 1978; and State of Maryland, Montgomery County, Department of Police, Office of Criminal Intelligence, Organized Crime Unit, "Re: Joseph Nesline," File No. OCR 75-12, 13 August 1978.

30. INTERPOL WIESBADEN, BUNDESKRIMINALAMT to Interpol Washington, "Investigation Reports submitted by the Landeskriminalamt Hamburg and Kiel, Re: Joseph Nesline," variously dated in 1977-78-79.

31. Interpol Teheran to Interpol Tokyo and copy to Interpol Washington-Rome, "Reference Radio-Message No: 386 of 6th June 78 regarding Dargahi Daviad dob. 5th Feb. 33 Iran," 11 September 1978.

32. Interpol Wiesbaden.

33. State of Connecticut, Department of State Police, Statewide Organized Crime Investigative Task Force, "Historical Data Sheet Re: Morris Joseph Salow," sent to Department of Police, Montgomery County, Maryland, 12 April 1978.

34. Harrison E. Salisbury, Without Fear Or Favor: The New York Times and its Times (Times Books, 1980). See particularly Chapter 25 "Traffic in Secrets."

35. Salisbury, 291-92.

Index

252

258